THE PERSONAL
ROBOT BOOK

To those robot experimenters, hobbyists, fans, enthusiasts, and writers who are succeeding beyond their wildest and most fantastic dreams in bringing the personal robot from the realm of science fiction to the level of fascinating reality.

THE PERSONAL
ROBOT BOOK

TEXE MARRS

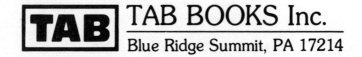
TAB BOOKS Inc.

Blue Ridge Summit, PA 17214

FIRST EDITION

FIRST PRINTING

Copyright © 1985 by Texe W. Marrs
Printed in the United States of America

Library of Congress Cataloging in Publication Data

Marrs, Texe W.
The personal robot book.

Includes index.
1. Robots. I. Title.
TJ211.M36   1985   643'.6   85-4630
ISBN 0-8306-0896-6
ISBN 0-8306-1896-1 (pbk.)

Cover photograph courtesy of Arthur P. Bond.

Contents

Acknowledgments

So many individuals, organizations, and companies assisted me in developing this book that it is impractical to list them all in this brief space. I owe a deep debt of gratitude to each of them, however, in particular the robot companies and hobbyists who generously provided me with a voluminous amount of information about their creations and products.

Appreciation is due also to those who furnished the photographs and illustrations that so amply showcase the fast-growing and incredible world of personal robots. In this regard, Marilyn Chartrand of Denver, Colorado and Sharon Smith, RB Robot Corporation, deserve special kudos.

To my wife, Wanda, a most heartfelt "Thank you!" As director of the consulting firm, Tech Trends, Inc., which she and I cofounded, Wanda is responsible for acquiring the library of information we maintain on high technology topics, including robotics. It was her research skills that made the writing of this book a most pleasurable and rewarding endeavor.

Introduction

In the latter half of the 20th century, we have been deluged by new technological discoveries. Every day brings word of a breakthrough in lasers, bioengineering, computers, space, or a dozen other burgeoning fields. Surely, however, the recent introduction of the personal robot into the homes of thousands of families ranks among the top technological achievements of all time.

True, today's personal robots are not yet up to the standards of R2D2, C-3PO, or the fascinating robot "people" envisioned by famed science fiction writer Isaac Asimov. In the short span of just a few years, however, scientists, engineers, and hundreds of robot hobbyists working out of their garage or workshop have been able to design incredible robots that walk, talk, smell, and remember. Some can fetch a canned drink from the refrigerator; others alert homeowners when fires start or burglars intrude.

The Personal Robot Book is the first comprehensive guide to these new and exciting companions of mankind. Here you can read of their dazzling capabilities and, of course, their limitations. Perhaps the perfect robot has not yet been created. Maybe he (or she) is now on the drawing boards at one of the many robot research laboratories now springing up around the globe. It could be that some Einstein-like robot scientist is at this moment formulating plans and ideas that may, before too long, produce a robot equal, or superior, to the human.

Still, for now, we can all be pleased and even thrilled at what is already available. Ours is the first generation to bring robots home.

Of what real use are these early robot creations? When we look to the skies and view the majesty of the Concorde or the Boeing 747, we harken back to the sage words of Benjamin Franklin. When he was in Paris to observe the experiments of the first hot air balloons, Franklin was asked by scoffers, "Of what use are they?" Replied the American statesman and scientist, "Of what use is a newborn baby?"

Part I

The Personal Robots Have Arrived

They Walk,
They Talk . . . They're Here!

The robots are not coming . . . they're here. They walk, they talk, they tell jokes. They are, unarguably, the most adored folk heroes of the 1980s—and they most certainly will change our lives.

At present, there are probably no more than 35,000 personal robots operating throughout the world, ranging in size from 6 inches to 1 feet. Eventually, most of us will have at least three of them in our home, predicts acclaimed robotics expert Robert Malone. "One will do household chores and work in the kitchen. Another will be an entertainment specialist like a cruise director, and a third will be a garage-based outdoor robot who rakes leaves and mows the lawn."

Of course, there are limits to just how fast this advanced technology will enter our homes. We may not, in the year 2000, come home to a robot-prepared sit-down dinner for eight—but there's no doubt that personal robots are making a slow but sure transition from fantasy to fact.

The robots have indeed arrived. Maybe they're not quite up to the talents of R2D2 and C-3PO, but their talents are considerable. Best of all, the robots are affordable and now interface with the personal computer. Personal robots are here-and-now creatures that provide companionship, educate youth and adults, play games and inspire fun, and perform innumerable household chores.

THE BUSY AND
ENTERTAINING PERSONAL ROBOT

This book will tell all about the new generation of robots, such as RB5X, which can vacuum the carpet for a busy housewife, a harried bachelor, or a career woman; B.O.B., which can fetch a beer for the exhausted manager who sinks into the sofa after a hard day at the office; and the Hot Tots, furry fantasy creatures that ride bicycles and perform all kinds of mayhem and antics to the delight of onlookers. The book will also discuss robots that can signal a warning in the wee hours of the morning if burglars intrude into the household, and robots that can detect and actually extinguish a fire.

There's much more, consider SMART RABBIT. This robot companion is very intelligent. He

OMNIBOT is just one of the many robots featured in this book. OMNIBOT has an on-board microprocessor, cassette tape deck, and remote-control transmitter (courtesy Tomy Corporation).

can carry on a sparkling conversation with the neighborhood kids, addressing each by name and confiding that he, too was once a "kid," though much smarter, of course, than people kids.

"But," a wise kid may retort, "you are controlled by a remote home computer. Ha! I've got my own brain!"

HERO I and HERO JR. help to teach the handicapped. Each has the patience of a saint, never becoming impatient with the learner. HERO JR. even cracks jokes. If the learner makes a mistake, he may offer encouragement: "That's close, but your electrical voltage is still a few watts shy!" Upon perceiving a correct reply, the approving robot responds with, "Bravo, you are an IBM-brain!"

OMNIBOT tells stories—amusing stories—

about his recent journey into space, where he encountered hostile alien beings. "What an adventure!" he exclaims.

And TOPO III? Well, TOPO III is a dancer—a pretty good dancer. Turn the stereo on and he'll dance to the music, singing a gleeful "wheee!" as he spins around at a frantic, fun-filled pace. Z-1 dances too: he wobbles, rocks to and fro, and lets out shrieks of joy, but only after beating his human opponent at one of his two favorite games: Blackjack and Tic Tac Toe.

THE AFFORDABLE PERSONAL ROBOT
Given today's computers and innovative software for robots, these electronic critters can be a regular member of the household. Imagine a friendly

household robot waking you up in the morning, newspaper in hand, with a cheery, "Good morning!" Then, at bedtime, consider the joyful faces of the kids when the robot tells them a bedtime story like that of Tom Sawyer and his whitewashed fence or the tale of Jack and the Beanstalk. Sounds like the 21st century has arrived a little early, doesn't it? But, all this fun and excitement is for real.

How much do you have to shell out to bring a new, robot family member home from the robot "pet store?" Surprisingly, the price tag for many of the newer robots is much less than you would imagine. For example, the price of many robots described in this book is under $100, though admittedly, the functions of low-cost robots are often limited. Consider, however, the fun and educational value of a robot like DROID-BUG, a Robot Shop creation. DROID-BUG will scoot across your living room floor and do all manner of stunts and tasks—all for an incredible $100. DROID-BUG and many other inexpensive, mobile robots talk, draw pictures, play games, and educate.

The prices of today's high-tech robots contrast sharply with the price tag of $15,000 for the personal robot offered by the Neiman-Marcus Department Store only a few years ago. The Neiman-Marcus model was featured as a gimmick for the store's fabulous Christmas catalog. Today's personal robots aren't gimmicks, and they cost appreciably less. Nevertheless, today's average robot is considerably more intelligent than the Neiman-Marcus model. Of course, you can still pay $15,000 and even more for your very own personal robot. Indeed, I'll be discussing custom-built robots which innovative companies like International Robotics and ShowAmerica can produce for you to meet your own dreams and expectations of what a robot should look like . . . and act like. Pack in powerful on-board computers and all manner of options and the price of a "Cadillac" model may go as high as $65,000, although $6,000 to $10,000 can also get you a fine robot.

HERO I and RB5X carry on a lively conversation at a robotics conference (courtesy Marilyn Chartrand).

ROBOTS: FROM DREAM TO REALITY

This book is much more than just a book about electronic and mechanical machines. It's a chronicle of man's supreme achievement to date: the creation of artificial life. True, today's robots are lacking in most respects when compared to the marvel that is the human body, but robotics is in its infancy and the future is bright. Why would man want to create a new breed to parallel himself?

Owning your own robot companion is a dream that man has seemingly held since the earliest times. Robots have been written about by writers, imagined by dreamers, and drawn and painted by artists for centuries.

While the march of the robot is definitely on, the concept of mechanization is not new. Robotics fascinated the collective consciousness as early as 3000 B.C. when man first conceived the use of strings and levers in constructing puppets and toys. Centuries later Aristotle wrote of a "mechanical man" who could work a loom, and Leonardo da Vinci drew sketches of mechanical robots. By the 1860s the French had invented spring-operated mechanical dolls, and Thomas Edison stepped into the picture with a talking doll he created in 1894.

As the twentieth century ushered in new levels of mechanization, robot technology moved briskly forward, but not, at first, without a skeptical reception. Drawing on a mistrust of machines at least as old as the Industrial Revolution a century earlier, the raging imagination of science fiction took hold. *R.U.R.*, Karel Capek's satirical play first produced in Prague in 1921, telescoped a horrifying panorama of World War I. Robots were depicted as alien beings created from a manufactured organic soup in a factory, and the play saw the first use of the word *"robot"* from *robata*, the Czech word meaning *forced labor*. A frightening vision of technology run amok became a favorite theme of the 1920s and 1930s, culminating in its most dramatic expression in the ravishing, but diabolical, robot Maria in Fritz Lang's 1926 classic film, *Metropolis*.

It took Isaac Asimov and his concept of the "civilized robot" to restore respectability to this mechanical species and quiet the fears of robophobia. In his classic 1949 book, *I, ROBOT*, Asimov set forth the "Three Laws of Robotics," three rules or principles that robots must be taught, or programmed, to always heed:

- ☐ A robot may not injure a human being or, through inaction, allow a human being to come to harm.
- ☐ A robot must obey the orders given it by human beings except where such order would conflict with the First Law.
- ☐ A robot must protect its own existence as long as such protection does not conflict with the First or Second Law.

As it became apparent that robots were not inherently evil—that their "thoughts" were simply the programs introduced into them—a more benevolent, and even whimsical, imagination began to shape the expression of robot fiction. Today, science fiction robots are sentimental and strongly emotional creations. They outfox us, as HAL tried to do on the fictional voyage to Jupiter in *2001*—or rescue us, as R2D2 and C-3PO did, from problems we can no longer solve and a high-tech future that grows dim and dark, where evil lurks.

One wonders what Karel Capek's reaction would have been had he known that 60 years after he first coined the word, personal robots would be mass-marketed, and exciting robot toys would line the shelves of stores around the world.

"The robot toy craze alone would stun Capek," said a spokesman for Tonka Toys recently, pointing out that Tonka's GoBots™ are selling like wildfire in toy and department stores.

"There's something magical about children's fascination with robotics," cays Robert Malone. "It's as if they really do sense that this is the future and they want very much to be part of it."

Robotics is, indeed, the wave of the future, and it is not just children who are fascinated by robots. A mysterious aura of high-tech wonder permeates this field as men, women, and children sense the fabulous challenge presented by robot technology.

THE ROAD TO THE INTELLIGENT ROBOT

The robots of science fiction, fantasy, and play are

Hasbro Industries offers "good guy" and "bad guy" toy robots. SOUND WAVE is an evil DECEPTICON™ robot warrier able to transform himself into an innocent-looking tape cassette player. His motto: "Cries and screams are music to my ears" (courtesy Hasbro Industries).

fascinating, but the story of real robots—intelligent machines—is equally fascinating. Ours is truly the first generation to experience the birth of a new life form, however primitive it may now be, and no tale in *Ripley's Believe It or Not* or in a science fiction thriller could be more exciting and breathtaking than the pleasure of taking an excursion down the robot road to the present.

Sowing Seeds for the Future

Until the twentieth century, the consensus of scientists and engineers was that an intelligent robot—an artificial person—was an impossibility. In 1883

Charles Babbage, a British inventor, conceived the "analytical engine," which some today credit as being the first computer. Babbage's device used punch cards borrowed from the Jacquard automated loom, an earlier invention. Babbage's machine was too crude to accomplish number-crunching or to have many industry applications. Certainly, no one at the time believed the analytical engine would be the basis for a future computer industry.

The 1930s and 1940s

The personal and industrial robots of today actually saw their genesis in the 1930s and 1940s as scien-

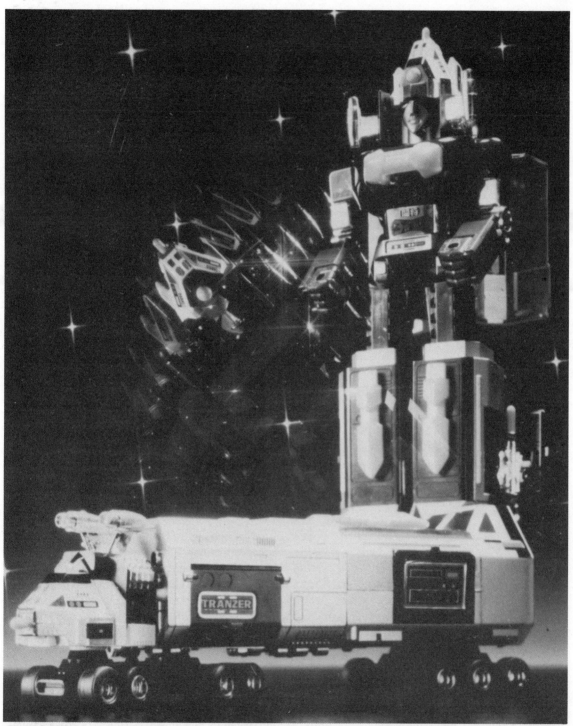

This is DAIMOS, a toy robot from Bandai America. With a few touches, twists, and turns, it is transformed into an amazing 20-wheeler vehicle (courtesy Bandai America, Inc.).

ABEGA is a highly detailed robot from Bandai America that transforms into a sleek, space-age craft (courtesy Bandai America, Inc.).

tists, engineers, inventors, and plain tinkerers, inspired by science fiction robots, began to apply their talents to building an artificial man. In these two decades, thrilled crowds poured into exhibition halls and world's fairs in Europe and America to get a glimpse of mechanical men on display. This was the age of electricity, and the bright hope was that, somehow, an infusion of volts and watts could inject life into these odd creatures.

One robot, "Elektro," was especially advanced. Built by a Westinghouse engineer, Elektro cavorted around at the New York World's Fair of 1939. He danced with beautiful women and generally excited admiring observers. Oh, yes . . . Elektro had a dog, Sparko, another Westinghouse creation, that actually stood on its hind legs and barked. Thus, the robot, which modern man depicts as some sort of high-tech pet and human companion, was depicted in Chicago as being a full person with a right to his own pet!

While visionary robot builders labored mightily to bring a workable machine to life, at the University of Pennsylvania, scientists funded by the United States Army were carefully building their own visionary machine: the first electronic computer. Dubbed ENIAC (Electron Numerical Integrator and Computer), the behemoth occupied the space of a two-car garage and contained over 18,000 vacuum tubes. By 1946, ENIAC was able to perform in a limited manner, though it was quite inept according to today's standards. Once, when the machine malfunctioned, researcher Grace Hopper found that bugs were shorting out electrical components. She told associates ENIAC had to be "de-bugged;" thus was born the term.

In the late 1940s, scientist William Shockley and others invented the transistor, a remarkable feat which made vacuum tubes obsolete. This led to today's integrated circuits and microchips and the miniaturization of the computer. Shockley's 1947 invention marked the beginning of the search for robotic "brains."

The 1950s

The 1950s found the show robot becoming more capable physically as a result of the use of more ef-

ficient, smaller motors and better designed mechanical systems. In this decade, some men began to conceive of the industrial robot. Engineer George Devol patented the first such device in 1954, and he and Joseph Engelberger founded the first industrial robot company, Unimation, in Connecticut. Then Engelberger succeeded in having a "dumb" but brawny and strong industrial robot installed on a General Motors' assembly line, and the wheel was set in motion for a fantastic change in humankind's manner of work.

Unimation is still in business, although the company is now a subsidiary of Westinghouse. Devol now owns his own consulting business in Florida, and Engelberger also is a well-paid consultant. Engelberger's book, *Robotics in Practice* (Amacom Books, 1980) is a classic guide to industrial robots for managers.

Also in the 1950s, computers began to grow more powerful, and the concepts of the intelligent machine and artificial intelligence began to take shape. In a particularly insightful comment that epitomized the growing recognition of "smart" machines and computers, Dr. Herbert Simon, noted computer scientist, stated in 1957:

"There are now in the world machines that think . . . learn, . . . and create. Moreover, their ability to do these things is going to increase rapidly until—in the visible future—the problems they can handle will be (those) to which the human mind has been applied."

The 1960s

The 1960s saw a number of important robotic advances. In 1962 the first robot was sent into space, and in 1968, SRI International, a California research group, presented Shakey to the world. Shakey was the first mobile robot with computer intelligence. He resembled an air conditioner on wheels. He was capable of avoiding obstacles and had a television camera, radio-link, range-finder, heavy-duty drive motors, and wheels. Finally, the world was made fully aware that the computer was destined to be the

Tonka Toys offers children its GoBots, mighty robots that turn into mighty vehicles. Also shown is the GoBot Commander Center (courtesy Tonka Toys).

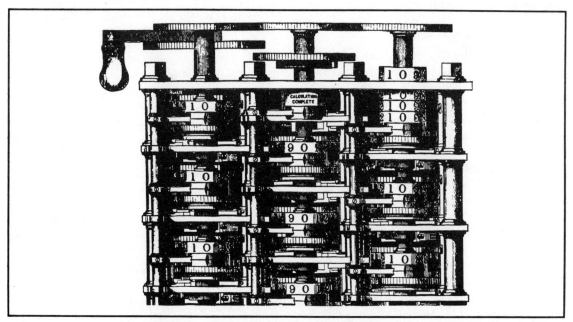

A portion of Charles Babbage's groundbreaking analytical engine (courtesy Pictorial Archives, Dover Publishing Company).

brain—the control center—for robots of all types. I can say with some degree of precision that 1968 with Shakey was the cornerstone for the age of personal robotics.

While Shakey was "shaking" man's conception of intelligence up a bit and proving the feasibility of the mobile, programmable robot, the computer front in the 1960s was heating up. IBM had taken a firm, early hold on the mainframe market, and giant computers began to proliferate. They were used primarily for large businesses and military and research applications, although in this decade it became fashionable to play computer chess opponents.

Conjecture by people with futuristic views was that someday a huge computer would be built which would control the world and become the master of mankind. In the movie *Dr. Strangelove*, the plot has the Russians in possession of a computerized retaliation "Doomsday" machine. Once the United States dropped the nuclear bomb on the Soviet Union, the computer would supposedly take over and direct Soviet missiles and bombers against America. Furthermore, the Doomsday machine could not be deterred from accomplishing its preprogrammed mission of destruction.

The 1970s

As the 1970s dawned, the Soviet Union followed the United States to the moon, but instead of cosmonauts, it was a Russian robot, the Lunakhod I lunar rover, that set "foot" on the faraway lunar surface. The robot scooped up rocks, took samples, made tests, and successfully made his appointed rounds.

In 1975, Apple Computer, Inc. launched the personal, or home, computer, following the microcomputer's commercial introduction by an earlier company. Apple, led by inventor-engineer and superpromoter/salesperson Steve Jobs, deserves credit for the computer revolution of the late 1970s. Apple made the industry what it is today. Reluctant IBM, seeing what a good thing personal computers could be to the corporate bank account, belatedly followed Apple into the sunlight of the microcomputer marketplace.

Also, later in this decade, the blockbuster movie *Star Wars* burst onto the scene, and every child and adult in America, it seems, began dreaming of someday owning their very own robot just like C-3PO and R2D2. Many robot hobbyists struck out to emulate the *Star Wars* duo, and a few entrepreneurs began to draw up plans for innovative new enterprises involving the production and sale of personal and home robots.

The 1980s: They're Here!

The personal robot burst onto the scene from 1981 to 1983 as several companies, including Heathkit, RB Robot, and Androbot, rushed their models to market. These were the first mass-produced home robots. Finally, the robots had come home. Families across the world could now hear the pitter-patter of tiny wheels treading across their living room carpet and view the happy faces of RB5X, TOPO, HERO, and their cousins.

In April 1984, an event of truly monumental importance took place. In Albuquerque, New Mexico, was held the first International Personal Robot Congress (IPRC). Assembled were all the known personal robots that could make the occasion (a few couldn't, presumably due to pressing business elsewhere,) plus founders of personal robot companies, robotics pioneers (such as Engelberger and others), hobbyists, inventors, and the plain curious. Isaac Asimov delivered the opening address via AT&T's Telstar 3 satellite, adding a futuristic note to the gala event.

The Golden Droid Award was given to innovative hobbyists. The most useful robot was adjudged to be the "RoboMower," an autonomous lawnmower. Later in this book, I will feature this device and its owner, David Falamak, head of EZ Mower, Inc.

The 1984 Albuquerque Congress was a signal to the world that the fledgling personal robot industry had arrived and should be taken seriously. It was a prelude to the Robotics Age.

THE FUTURE OF PERSONAL ROBOTS

The inventors of the new personal robots and the

entrepreneurs of the first companies mass-producing them realize that these electromechanical creations have far to go to reach their full potential. The future of personal robots will be a time of experimentation and continuous upgrading, punctuated by remarkable new discoveries and breakthroughs in mobility, artificial intelligence, and usefulness.

According to Future Computing, Inc., a Richardson, Texas market forecasting firm, the personal robot industry will emerge into a $2.2 billion industry by the end of the 1980s. During the 1990s, says this forecast, personal robots will be so functional they will have a strong impact on employment and labor trends. Furthermore, the economies of scale in manufacturing will allow mass-produced personal robots to provide greater capabilities at lesser cost than today's models.

Nelson B. Winkless, a well-known robotics authority, addressed the subject of the future of personal robots at Albuquerque's International Personal Robot Congress in 1984. Winkless believes that the future will see a robot home appliance that vacuums, dusts, and feeds the family pets. The robot will place a "hand" on your knee, be able to sense if your blood pressure is a little high, and recommend a visit from the medical doctor. When the doctor arrives, it will be the personal robot he talks to because the robot stores the whole family's vital health statistics.

Many other experts predict a profound change in our lifestyles due to technological advances in personal robotics. Robert Malone, noted robotics author, says robots could be given the personality of a Cheshire cat or an angry dog. They can also be decorated to fit the decor of a home. Nolan Bushnell, founder of Atari and Androbot, would agree. He states that in the year 1990, "Over one million home robots will be purchased each year by American consumers."

Patrick H. Winston, director of the Massachusetts Institute of Technology's Artificial Intelligence Laboratory, affirms that advances in artificial intelligence will very soon allow computers and robots to reason, learn from experience, and communicate easily in human language. Winston

emphasizes that this future may be a ways down the road, but it nevertheless is inevitable. Of today's era, Winston remarks, "It's like being at Kitty hawk when the Wrigth brothers' plane took off."

The analogy of the personal computer industry may be revealing. When personal computers were first offered in the mid-1970s, few foresaw millions of the machines invading homes within a few short years. That pattern could repeat itself in the personal robot field. Carl Helmers, publisher of the widely read and respected *Robotics Age* magazine, suggested this could be the case. Said Helmers, "The Hero I (personal robot) marks a turning point comparable to the introduction of personal computers in kit form in 1975."

The Ultimate Personal Robot

Increasingly, the science of *bionics*, or *biomedical engineering*, is giving humankind spare parts medicine—artificial body parts made up of materials compatible with biological tissue and patterned after actual human physiological systems. The American Society of Mechanical Engineers furnished the following article that examines the future possibility of a robot closely patterned after the "ultimate machine"—the human body:

A machine that runs on a wide variety of fuels, emits only recyclable or biodegradable materials, accepts a wide range of informational inputs, performs amazing feats of data processing, and produces a wide variety of output functions may just be the ultimate machine, according to George Piotrowski, associate professor, Department of Mechanical Engineering, University of Florida, Gainsville. This machine is the human body.

Although mechanical engineers have been applying engineering principles to biological systems, Pitotrowski suggests in an article in the March issue of *Mechanical Engineering Magazine*, a publication of the American Society of Mechanical Engineers, that engineers apply the biological principles and devices found in the human body to their design strategies.

Human bearings—joints—allow motion with a minimal loss of energy to friction. The friction reported in some joints is lower than that found when using Teflon. Also, there is normally no wear seen on the surface of human joints. Bearing designers might consider the model of human joints, which sustain loading of 3 to 10 times body weight, when designing new bearings.

The human skeleton, with its bones, tendons, cartilage and muscles is an excellent study in weight bearing design. Certainly, it would be difficult for a designer to mimic the self-repairing quality of human bone, but other aspects of bone could be advantageous in the design process.

"The mechanical structure of the body is built up of optimally designed links made of a fatigue-resistant composite material," Piotrowski stated. Human bone employs dense material in a hollow tube fashion for weight bearing. Spongy, more resilient bone on the ends of long bones make them more resistant to impact injuries.

One of the most fascinating aspects of the skeleton is the way the muscles, tendons, cartilage and bone work together to strengthen the structure. The attachment of tendons to bone is gradual, with bone turning little by little to the collagen of tendons. This eliminates the stress that would occur if the two dissimilar components were simply attached.

When it comes to data processing, the human brain and nervous system is an excellent example to follow. Localized control, such as that found in the knee-jerk reflex, avoids central processing completely, while information flow back and forth between the brain and the areas of the body along the nerves is "remarkably immune to noise and has high reliability," according to Piotrowski.

"Nature in all its aspects . . . provides the engineer with myriad examples of sophisticated designs," Piotrowski said. "Their study will only enhance the designer's creativity."

The Robot Anatomy:
Obstacles to Full Development

There are still several physical obstacles that must be hurdled before the personal robot takes its place beside the human as a full and almost equal partner. Let's briefly examine the state of the art and project the future direction of anatomical change for personal robots.

Vision. Robots are now limited in sight. Today's models use a TV camera as a sensor, with the camera translating what it sees into pixels. Each pixel is assigned a numerical value based on the light emitted and observed. The numerical values are patterned by the computer and analyzed against the computer's memory bank. The robot can then respond appropriately to what is "seen." For example, it can avoid an obstacle or pick up a part.

Vision sensors are fast becoming super-sophisticated. Today's robots could legally be classified as blind, but tomorrow's may see as well as many vertebrate animals. According to *Omni* magazine (July, 1983), an engineer and an entomologist are now working on a robot eye that may someday be superior to the human eye. Patterned after the eye of the horsefly, this new eye being researched resembles the ball end of a microphone, so it is able to capture 360 degrees of a scene, whereas the human eye views considerably less. The eye senses light that is sent to a computer by fiber optics. The computer then analyzes the picture it sees.

Artificial Intelligence. The horsefly eye just discussed, to be truly workable, needs a computer with advanced artificial intelligence to analyze what is seen, make judgments, and act accordingly. Current computers are limited, but much work is being done. Many scientists are studying the human brain and sensory system for clues to a more effective artificial intelligence system. Herbert Simon, of Carnegie Mellon University, says that today's robots don't have a lot "between the ears." The ultimate solution, say some, is a molecular, or biological chip. Such chips may be available by 2000 A.D.

Tactile Sense. Tactile sense (touch) is now limited to perception of size, temperature, and to a small extent, texture. Future sensors will be better able to determine texture and form, and adjudge material makeup of an object; for example, whether it is glass, wood, or metal and whether it is smooth or grainy. Robot hands now aren't adaptable for most human needs; no robot invented can open a refrigerator door, grab a can of soft drink, and pop the top. Such nimble capability is probably more than a decade away.

Olfactory Sense. The best of future personal robots would have an olfactory (small) sense approximating the facility of the human nose. Robot noses are becoming sophisticated. Some can now smell toxic gases, for example. Scientists say that bioengineering methods will produce a keen sense of smell for robots within a decade. Future biosensors will be able to detect one molecule out of a billion, say University of Toronto Chemists Michael Thompson and Ulrich Krull. Who knows, the future home robot may even perform the odious task of informing a visitor that he or she should change deodorants.

Mobility. Robots need better locomotion systems. New prototypes walk with multiple legs, much as spiders do. One bounces like a pogo stick. Future models will be two-legged. Look for them by about 1995. Mobile robots with flexible hands will give us the lawn-mowing, window-washing robots we all desire.

Language. The old Votrax voice system with 64 phenomes and its machinelike, stilted sounds is giving way to speech systems which are remarkably similar to human voices. Computerized robots today, however, cannot communicate without humans employing complex programming languages. Natural languages will be developed and coupled with artificial intelligence so that by the year 2000, your home robot may be able to intelligently discuss with you the merits of, say, the two presidential candidates.

WHAT IS A ROBOT?

It seems foolish to ask the question, "What is a robot?" The many varieties and types of robots, however, make the question worthy enough to ponder. The fact is there are mobile robots and stationary robots; tall robots and short robots; "smart" robots with microcomputer brains, and "dumb" robots that can only perform a single, limited task. Some robots have flexible arms with semidexterous claws, or hands; others are armless. There are robots that weigh tons and can pick up a truck body, and there

are robots that strain to lift 16 ounces. A few robots can "see;" most are blind. Interestingly, the robot population is even more varied and wide-ranging in performance and physique than the human population, with its own vast range of colors, shapes, sizes, abilities and temperaments.

What, then, is a robot? To most of us, a *robot* is simply an artificial creation, usually with electrical and mechanical parts, that mimics the acts, or behavior, of humans. That's basically the definition furnished by one of the greatest of robot "forefathers," Joseph Engelberger. In *Robotics Age* magazine (April, 1984), Engelberger considered the distinction between hobby, or personal, robots and industrial robots and sagely concluded that, "A robot is a robot is a robot." Said the distinguished engineer, credited with introducing the industrial robot to America:

"Industrial robots are gaining sensory perception, advanced language and mobility. And so are the hobby robots. With so many bright enthusiasts in the game the two robotics cultures must converge. Before this decade is out, it should become evident that robots no longer need the distinctions *industrial* and *personal*. They will all be just robots "under the skin."

Engelberger's definition is about the best I have seen. There are others, of course. People love to come up with definitions. According to some experts, a robot just must have a computer on-board. For others, any machine controlled by a computer—on-board or remote—can properly be classified as a robot. Then, there are those who so broadly define the term "robot" that an automobile, cow-milking machine, or electric coffeemaker fit the definition.

The Personal Robot As Space-Age Pet

Whatever definition you choose, when you consider the term *personal robot*, your thinking may change. A personal robot (also called *hobby* or *home robot*) conveys a creation that is near human. Not necessarily in human form as we envision androids and humanoids, a personal robot is still expected to be embodied with features or characteristics that lend it compatibility with the human family. Some would suggest that the personal robot must be a companion, an extension of a person, in other words, a sort of space-age pet.

Until such time that robots are endowed with fifth-generation computer brains so that they think and feel as humans do, they are in essence exactly that: space-age pets.

The personal robot industry is a historical phenomenon. Indeed, this book is itself a history book, in that it presents a whopping piece of man's technological history—the creation of a new breed on earth. What a remarkable future lies ahead as personal robots graduate from the status of a hobbyist toy to space-age pet and, finally, take their place as man's full-fledged partners.

The Things That Robots Do

Perhaps the best way to understand what a robot is, is to look at what it does. For example, over the next few pages you will find photographs of several types of robots, including:

☐ A mailmobile delivery vehicle robot from Bell & Howell preprogrammed to independently ferry mail and internal distribution within one floor of an office building.

☐ The Space Shuttle's robot arm, a tool that has proven invaluable to astronauts in retrieving and repairing errant or immobile satellites.

☐ Arm robots. For example, an ASEA industrial arm robot that assists in the

An example of a tabletop robotic arm is this device produced by Cyber Robotics of Cambridge, England. Computer-controlled, such arms are excellent educational tools (courtesy Marilyn Chartrand).

16

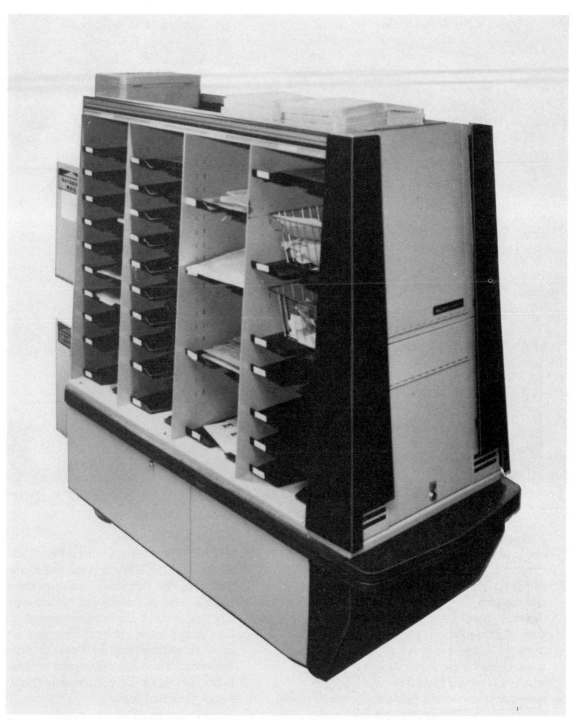

The Mailmobile® , a self-propelled delivery vehicle from Bell & Howell (courtesy Bell & Howell).

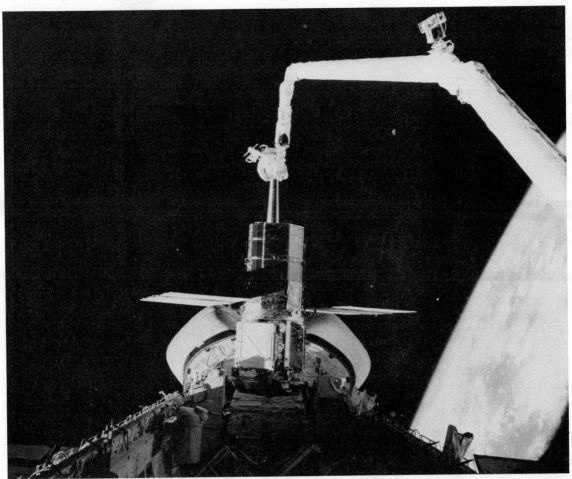

Space missions would be severely limited in work without the Space Shuttle's robotic arm (courtesy Lyndon B. Johnson Space Center--NASA).

automotive industry by applying urethane sealer to windshields. Also, note the less-expensive assembly robot, Hitachi's A4010H, and Cyber Robotics' small tabletop arm.

☐ "Robot Redford," a promotional, or show, robot that gained fame of sorts by becoming the first robot ever to speak at commencement exercises for Anne Arundel Community College in Maryland.

☐ Two other promo robots: SICO, a robot that travels the world handing out business cards and promoting products, and an honest-to-goodness, walking, talking Heinz tomato ketchup robot.

☐ CHARLEY, a hobbyist's robot *par excellence,* the work of Jim Hill of Covina, California.

☐ ODEX I, an incredible walking robot manufactured by Odetics, Inc. in California. The huge ODEX I is powerful enough to lift a pickup truck, yet it can squeeze itself into a shape small enough to pass through a doorway. This robot is designed to assist in fire fighting and for military, security, and heavy industrial uses.

☐ The Digirobo Tokima robot watch, distributed by Bandai America. This unique little fun gadget is a toy, yet . . . it's more. A digital wrist watch, it converts into a tiny free-standing robot clock. You know the old saying, "What will they think of next?"

As you can see, robots have many functions. Their versatility and the fact that they can be transformed as the need arises into so many shapes makes them very valuable members of the human team. In a sense, all robots are "personal," but whether robots are personal or impersonal, they're here to stay.

A ROBOT COMPANION: THE ULTIMATE COMPUTER PERIPHERAL

Many personal robots are controlled by computers, either remote or on-board; thus, there is a "computer-robot connection." The most sophisticated personal robot is an extension of the computer. In fact, the robot is the ultimate computer peripheral. A good, concise description of the computer-controlled personal robot is that he, or she, is a computer with humanlike appendages and traits. The robot, with its computer brain, is perhaps the greatest achievement of technology. It perfectly complements the imperfect computer. Lacking mobility, sight, touch, and other human senses, the computer is limited to intellectual accomplishment. Give the computer these senses through a link-up with a robot, and its capabilities become limitless.

The Personal Robot-Computer Connection

Personal robots are extremely attractive to the

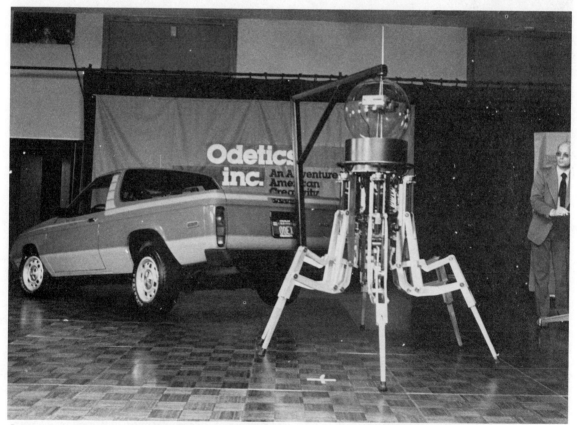

Odetics, Inc., the maker of ODEX I, calls it the world's first "Functionoid." The robot combines state-of-the-art engineering, electronics, and computer science (courtesy Odetics, Inc.).

This heavy-duty industrial robot toils at the Windsor, Ontario Chrysler assembly plant applying urethane sealer to car windshields (courtesy Chrysler Corporation).

millions of personal computer owners who are constantly searching for ways to enhance the effectiveness of their computers and upgrade their capabilities. The peripheral market is robust. The experts say that for every $1 spent on the computer itself, $3 or $4 are spent on accessories and peripherals. The exponential growth of the software market is evidence of the willingness of computer owners to discover new uses for their computer. The proliferation of personal robot models, many priced less than $1,000, add a new dimension to computer interfacing.

Many of the personal robots described and illustrated in this book are compatible with the most popular brands of personal computers. Quite a few have computers of their own, on-board. BASIC,

Logo, and Forth are languages used to program robots. By interfacing the robot with their computer, owners are putting arms and legs on their computers.

Imperfect Interfacing

This is not to say, however, that the robot and computer can be perfectly interfaced. No way. Problems—severe problems—exist. For one thing, programming a computer to get a personal robot to correctly accomplish even the smallest of tasks (such as lift its arm to a certain level) can take hours of exhausting programming time. There's also the compatibility problem. Just as system standards vary among different types and models of computers, so do standards vary among robots, many

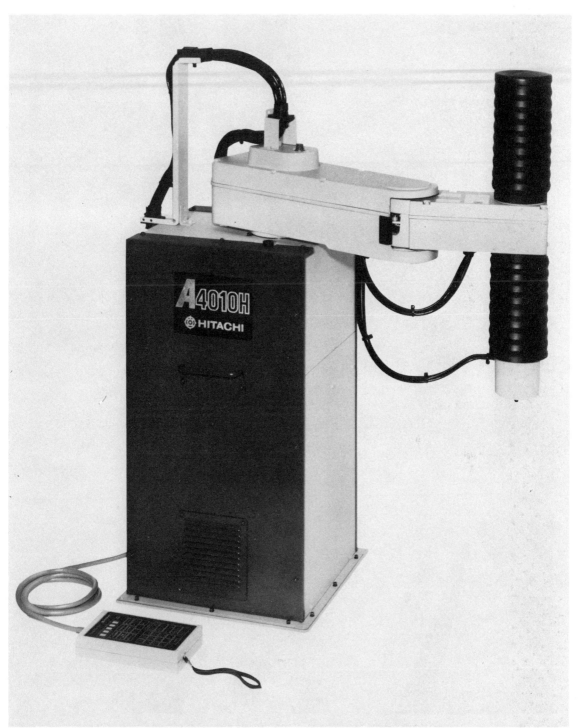

The Hitachi model A4010H Assembly Robot is low-cost (under $10,000) and lightweight, and has a resident memory of 200 steps. It is designed for both mechanical and electronic assembly (courtesy Hitachi).

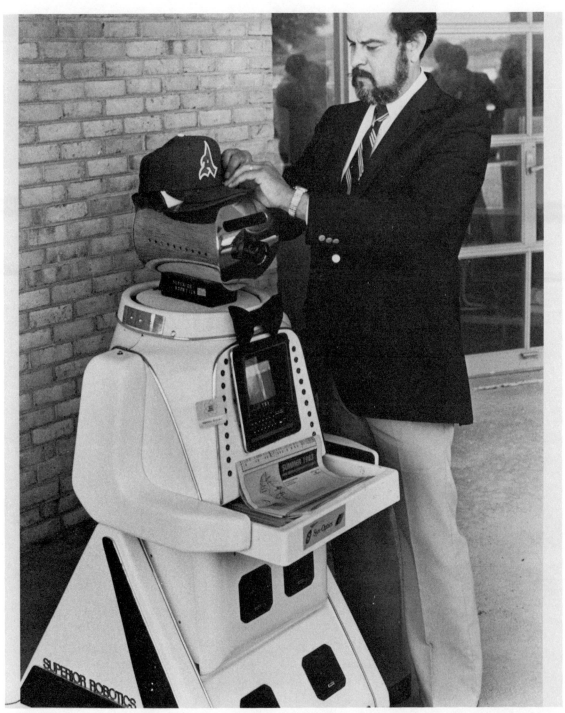

"Robot Redford," a promotional robot owned by Superior Robotics, was a speaker at commencement exercises at Maryland's Anne Arundel Community College. In appreciation a college official presents a cap with the letter "A" to the handsome robot (photo by Rob Hendry, courtesy Anne Arundel Community College).

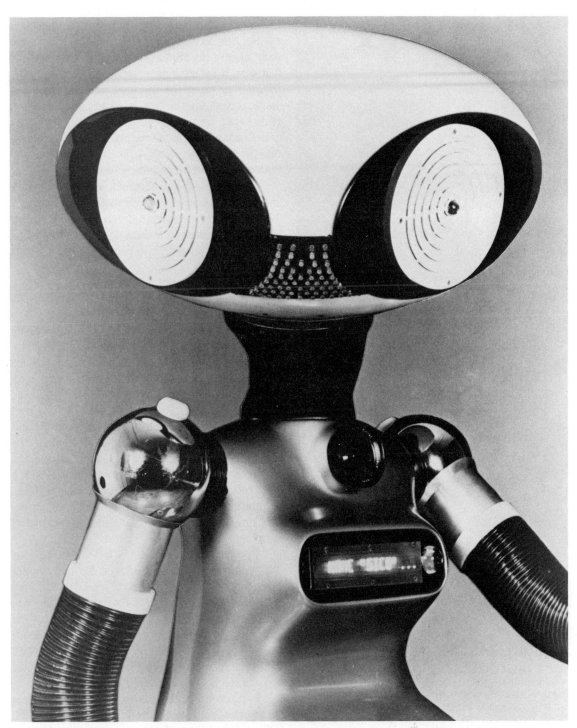

SICO is his name and attracting crowds is his game. You'll read about SICO and his friends later (courtesy International Robotics, Inc.).

"HJ" is a 6-foot-tall walking, talking ketchup bottle robot which represents the H.J. Heinz Company at special events, ranging from trade shows to state fairs. "HJ" is one of many robot product replicas built by ShowAmerica Inc., Elmhurst, Illinois, the world's oldest and largest promotional robot company (courtesy ShowAmerica, Inc.).

of which possess contrasting operating systems.

So, keep in mind that even though a robot company claims its product can perform miraculous feats, such as fetching a beverage can from the fridge, delivering the morning newspaper with a smile, or walking the dog, keep an open mind and realize that, well . . . robots are nice, but they're not quite human—yet. Brain (i.e., computer) limitations may prevent your robot from performing as the high-tech wonder it is touted to be.

Another limitation often lies in robotic software, or the lack thereof. Many robots come complete with handy software programs on disk. Also, several independent software makers now offer software for specific robot models (see Part III, *The Hobbyist's Corner*).

Sample Programs

Whereas programming a robot yourself can be an arduous task, running an established computer program for a personal robot can be a breeze. For example, the ComRo ToT personal robot comes with these taped programs:

Program #1. Press **GO C40 CR** (and turn speech on). When someone passes in front of ToT, his eye will light and he will say: "Welcome to the Tack Room." This is a good stationary program in which there is no base movement. ToT will wait a few seconds before repeating the message. If you are going to put ToT up on a table or stand, disconnect one lead from each base motor just to play it safe.

Program #2. Press **GO C80 CR** (First raise ToT's right arm perpendicular to his body under radio control and plug in his tray module with some snacks or hors d'oeuvres.) ToT will proceed forward until he comes to a person (or thing), then say: "Excuse me . . . may I serve you?" Then he passes for the person to take something, says, "Thank you," backs up, turns, and goes on to another person. He repeats his actions until you press Reset.

Program #3. This is an interactive program that works best with children. Press **GO CEO CR**. ToT advances to the first person or thing. His eye

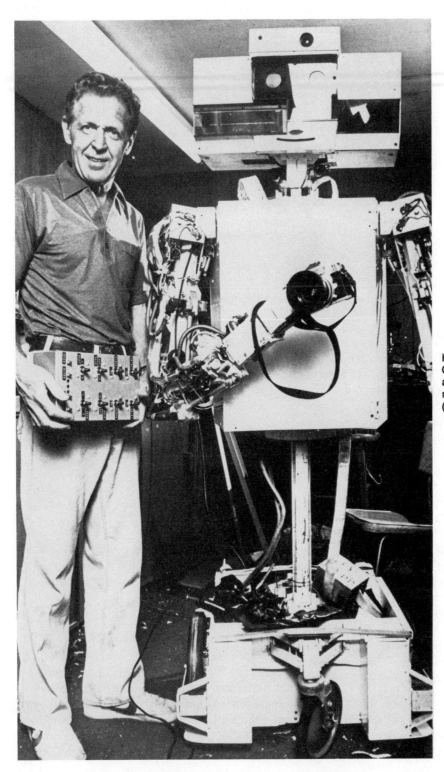

Hobbyist Jim Hill of Covina, California, has won much acclaim for his remarkable robot invention, Charley (courtesy Jim Hill).

The Digirobo Tokima wrist watch, from Bandai America, becomes a free-standing robot within seconds (courtesy Bandai America, Inc.).

lights. He backs up. He says, "What's your name?" He puts out his right hand and shakes, saying, "I'm TóT, nice to meet you. Welcome to the Tack Room." Then he advances, turns right and moves off to the next person, stops, says, "Hello, big guy . . . What's your name? . . . That's correct . . ."Shakes. Says "Good to see you." Turns right. Advances. Says: "Hi there . . . Who are you? I am TóT . . . (shakes) Welcome to the Tack Room." Advances, turns right. At this point TóT pauses and watches for motion. If he sees nothing he stays still. If there is motion in front of him, he starts the program from the beginning.

Children love shaking hands with TóT. If you keep the arms a little bit loose at the shaft coupling to the shoulder, they can't do much harm, as TóT is quite rugged.

Another Sample Program. Here is another sample program, this one written in programming language. It is a program for Androbot's TOPO, written in TOPOBASIC. This program makes TOPO walk in a square.

```
5    REM SET-UP LOOP
10   FOR J = 1 TO 4
15   REM COMMAND = FORWARD,
     DISTANCE = 100
```

```
20   C = F : P1% = 100
25   REM EXECUTE THE
     COMMAND
30   GOSUB 2
35   REM COMMAND = LEFT,
     ANGLE = 90
40   C = L : P1% = 90
45   REM EXECUTE THE
     COMMAND
50   GOSUB 2
55   REM END OF LOOP
60   NEXT J
```

(GOSUB 2 sends you to a subroutine that is built into TOPO BASIC. Thus, there is no line 2.)

A ROBOT GLOSSARY

As you proceed through this book, and especially as you read Part II, which describes the workings and performance of the various models of personal robots, you may find some of the terms, acronyms, and abbreviations unfamiliar. In the back of this book is a Robot Glossary. Turn to that section when you encounter a word alien to your vocabulary.

Part II
Robots Galore:
A Gallery of Personal Robots

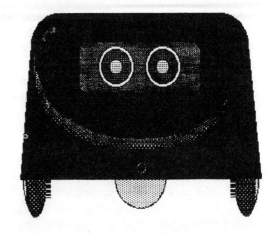

Robots Galore:

In the following pages, you'll meet virtually every personal robot on the market today. The introduction of workable personal robots has created a multitude of intriguing questions. In reviewing the world of functional robots, here are some of the questions I have attempted to answer:

- ☐ How does this robot work?
- ☐ What are its capabilities? Can it walk, talk, see, hear, feel?
- ☐ What is its "brain" capacity (microprocessor characteristics)?
- ☐ With which home computers is it compatible? (Apple, Atari, Commodore, TI 99/4A, IBM PC, etc.)
- ☐ Can it operate independently without interfacing with a computer?
- ☐ What accessories and software come with the package?

A few caveats apply. In considering these points, note how related circumstances are in the computer and robot industries.

- ☐ *About the Dynamic Robot Market:* The prices, descriptions, and specifications for the robots, options, and accessories listed here were provided by the manufacturers, distributors, and/or other authoritative sources. Needless to say, they are subject to change at any time. The robot market is most dynamic.
- ☐ *Buyer Take Care:* I have probably seen more personal robot demonstrations than any man on earth (*Guiness Book of World Records,* are you listening?) Nevertheless, not all of the robots listed have been seen and personally reviewed. The robot companies have been generous in providing information about their products, but it is quite possible that some of their performance information is self-serving. My advice: before you plunk down money for a robot, write or phone and acquire as much information as you can about the product and the company. If possible, see the robot demonstrated.

- Some of the robot companies listed here and their products may not survive. There is an ongoing industry shakeout. Also, as in the computer industry, promises of future enhancements or of options to come may, in some instances, prove groundless, even though unintentional. Perhaps research snags may occur, or the company may fall on hard times and delay or cancel introduction of the promised refinements. Again, buyer, take care.

- In reading of a particular robot's abilities to perform complex tasks, such as extinguishing fires or vacuuming the carpet, keep in mind that the capability in this regard may be of a rudimentary level. This is not to say that robots are nonfunctional; their capabilities are often surprising and occasionally remarkable. Personal robots are, however, better at some tasks than others, which is another way of suggesting that robotics is still at the infancy stage.

Now . . . on to the greatest gallery of robots this side of the Milky Way. In the following pages, company by company, I review the personal robots offered by America's fledgling robot industry, as well as a few from overseas. The United States is at the moment, however, the hotbed of personal robot research and manufacturing. Addresses and phone numbers are found in Appendix B.

HEATHKIT

The best thing about the Heathkit robots HERO I and HERO JR. is that they're made and marketed by Heath Company, one of America's most reliable electronics manufacturers. Heath was one of the first to offer personal computer kits for hobbyists. That was in the days before companies like Atari, Texas Instruments, Commodore, and Apple jumped headfirst into the personal computer market. For 35 years, Heath has been a direct mail marketer of electronic products and do-it-yourself kits. In 1979, Zenith Electronics Corporation purchased the Heath Company, and a subsidiary, Zenith Data Systems, took charge of Heath's operations.

In 1983, Heath (popularly called Heathkit) introduced HERO I, one of the first personal robots. Actually, it was designed and built for educational use, to teach individuals robotics and industrial electronics. The robot has proven so successful that in late 1984, the company came out with HERO JR., designed primarily as a home companion robot. The HERO I must be programmed by the owner, although a modification addition and limited, off-the-shelf software available from a few companies can make the robot operable from a remote home computer. Unlike its "big brother," however, HERO Jr. is preprogrammed to carry out its activities and functions. See Table 2-1.

HERO I has sold better than any other personal robot—about 15,000 units so far. The Heathkit reputation should also enhance sales of HERO JR. Because of this assured market, these two able and popular personal robots deserve special attention in this book. Let's start by examining the older of the two: HERO I.

HERO I: The Educator

HERO I is an intelligent, mobile robot designed to teach the principles of robotics to the builder and user. It has features and capabilities not found on industrial robots costing many thousands of dollars more.

Physically reminiscent of the robot R2D2 of *Star Wars* fame, HERO I stands 20 inches tall and weighs 39 pounds. Its most prominent feature is its turretlike head, which rotates up to 350 degrees and carries its arm mechanism, programming keyboard, and experimental circuit board.

It is a completely self-contained, electromechanical robot capable of interacting with its environment. It is controlled by its own on-board, programmable computer and carries electronic sensors to detect light, sound, motion, and obstructions in its path. It can be programmed to pick up small objects with its arm, speak complete words and sentences with its voice synthesizer, travel over predetermined courses, and repeat specific functions on a predetermined schedule. It carries its own rechargeable power supply and can function totally free of any external control.

Table 2-1. A Concise Look at the Heathkit Robots.

HERO 1	HERO JR.
ROBOT TYPE: Education Trainer Robot	ROBOT TYPE: Personal Robot
MARKET: Schools, Industry (Training), Individuals (Home Study)	MARKET: Consumer, Home
USES: Robotics Education/Experimentation	USES: Friend, Companion and Security Guard
CAPABILITIES: Robotics and industrial electronics trainer. Intelligent, mobile robot with on-board programmable computer. Using sensors, it can be programmed to interact with its environment. Five-axis arm allows manipulation of objects.	CAPABILITIES: Fully pre-programmed mobile robot. Speaks, sings songs, recites poems and nursery rhymes and plays games. Also acts as security guard, alarm clock and date reminder.
SENSES: Light, sound and motion detectors plus sonar.	SENSES: Light, sound and motion detectors plus sonar.
PROGRAMMED BY: RS-232 computer interface, teaching pendant, RF remote control, cassette interface and hexadecimal keypad (on-board).	PROGRAMMED BY: Fully pre-programmed. Additional pre-programmed game/entertainment cartridges available. RS-232 computer interface. (HERO JR. *BASIC* cartridge requires terminal or computer with RS-232).
ACCESSORIES: Wireless remote control. RS-232 Computer Interface. Memory and Program Chips. Two courses in robotics and industrial electronics and one course on advanced robot programming. Various demonstration programs.	ACCESSORIES: Infrared motion detector. Wireless Remote Control. Optional pre-programmed game/entertainment cartridges.

Heathkit/Zenith Educational Systems produce a complete educational course on robotics which can be used in conjunction with HERO I.

In the words of Douglas Bonham, director of Heathkit/Zenith Educational Systems, "the Heath robot incorporates all of the basic systems found on modern industrial robots, plus a few that are still in the experimental stage of industrial application.

"The Heath robot makes it possible for nearly anyone to obtain a comprehensive education in robotics," Bonham says. "It also puts effective robotics training well within the financial reach of individuals, schools and businesses."

The complete HERO I kit, which includes everything needed to build the computer-controlled,

multifunction robot, is priced at just under $1,500.00. The robot may also be purchased fully assembled for $2,200.00.

The Versatile HERO I. HERO I incorporates a host of features that make it an amazingly versatile device. Its on-board computer can be programmed to guide the robot through various complex maneuvers, activate the robot's sensors, and modify the robot's behavior in response to inputs from its on-board sensors and real-time clock.

HERO I can be programmed with either its on-board keyboard or the external training pendant supplied with the unit. Programming can be stored on any standard audio cassette tape recorder and reloaded for future use. An optional RS-232C feature permits programming from a host computer.

In the process of working with HERO I, the owner or operator gains practical, hands-on experience with the basic elements of robotics, such as programming, electronically controlled drive and positioning, interfacing, and data acquiring. HERO I isn't limited to teaching robotics, however. The robot also provides useful exposure to a number of other, related disciplines that have broad industrial applications, such as industrial electronics, automated equipment, electromechanical devices, and programmable controllers.

The Heath HERO I robot also features an experimental circuit board which lets the operator interface circuits of his own design with the robot's computer.

"Within the parameters of its basic physical limitations, there is virtually no limit to what you can do with this robot," says Bonham. "It really becomes a question of just how much skill, imagination and ingenuity the operator has."

A three-wheeled base with both drive and steering on one wheel propels the robot in any direction. Its 12-inch minimum turning radius gives the robot exceptional maneuverability. It carries on-board sensors to detect movement, sound, and light, plus an ultrasonic ranging system to detect objects and obstructions in its path. The robot also has an arm and gripper mechanism with which it can pick up small objects. It can also be programmed to speak complete words and sentences with its speech synthesizer.

The robot's computer can be programmed three different ways:

- [ ] Through the keyboard mounted on the robot's head;
- [ ] with its handheld remote control teaching pendant, or
- [ ] through its cassette serial port using programming previously stored on a conventional audio cassette tape recorder.

The computer is capable of storing programs with over 1000 individual steps.

The robot's eight motors and logic system are powered by four rechargeable batteries, which are protected against total discharge by an automatic low-voltage sensor. The HERO I Robot comes equipped with its own external battery charger and can be operated while the batteries are being recharged.

The HERO I Kit. Heath's HERO I Robot consists of three separate kits:

- [ ] The basic robot kit, which includes everything except the arm and voice synthesizer, and sells separately for $999.95;
- [ ] The arm mechanism, which sells separately for $399.95;
- [ ] The speech synthesizer, which sells separately for $149.95.

The basic robot is fully functional and can be purchased separately. Both the arm and the speech synthesizer can be ordered as accessories. Purchasers ordering the total package of basic robot, arm mechanism, and speech synthesizer will pay just under $1,500.00, plus shipping charges and tax, a savings of about $50.00 from the total cost of the three kits purchased separately.

Heath officials describe the HERO I Robot kit as one of the more challenging kits the company produces. It is not for novice kit builders, they say. Builders should have some experience in kit building and electronics, including circuit board building, soldering, and wiring harness assembly.

Simple Heathkit instructions guide the builder through each assembly step, using parts supplied in the kit. No special tools are required. Heath estimates it will take the experienced kit builder 40 to 60 hours to assemble the complete robot, including the arm mechanism and speech synthesizer.

Specifications for HERO I. The Heathkit robot is a state-of-the-art product, incorporating the same types of systems found on today's most current industrial robots.

Sound: The robot's sound sensor detects and quantifies ambient sound levels over the frequency range of 200 to 5000 Hz. The sound sensor is approximately omnidirectional and, like all sensors on

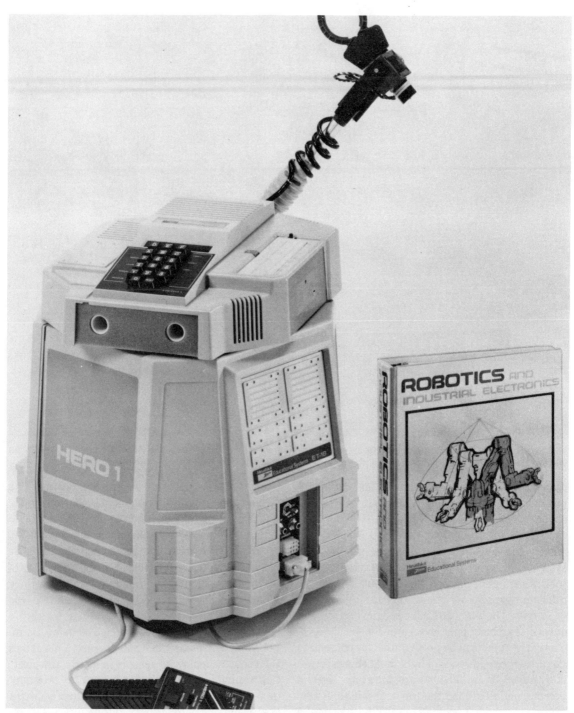

The Heath HERO I Robot (Model ET-18) is one of the most advanced and versatile products for teaching and learning robotics on the market today. Its companion Robotics Education Course provides a comprehensive grounding in robot technology (courtesy Heath/Zenith).

The Heath HERO I Robot Do-It-Yourself Kit is the most unique kit ever introduced by the Heath Company. It combines electronics, mechanics, and microcomputers into a totally new kit-building experience (courtesy Heath/Zenith).

the unit, can be interfaced with the on-board computer to allow the robot to respond to external sounds.

Ultrasonic Ranging: The robot has its own pulsed ultrasonic system with a range of 8 feet. The system can discern objects and obstructions in the robot's path and, through the computer, warn the robot to take appropriate action.

Light: The on-board light sensor detects and quantifies ambient light levels over the visible spectrum. The robot can be programmed to respond to the presence or absence of light in its environment.

Motion: A continuous-wave ultrasonic system detects motion around the robot. It will detect an average-sized adult walking toward the sensor at a distance of up to 15 feet.

Speech: The robot can be programmed to talk using its phoneme-based speech synthesizer. The system can generate 64 phonemes (basic sounds) which can be linked together in any combination to simulate human speech and various sound effects.

Time: The robot has its own 4-year calendar clock which counts seconds, minutes, hours, day of week, day of month, and month of year. The clock allows the robot to record and respond to the passage of time.

Head: The robot's head, which contains the robot's four sensors, rotates 350 degrees. This allows the robot to deploy its sensors in nearly any direction around it. The head also houses the programming keyboard, LED display, experimental circuit board, and one of the robot's four batteries. The robot's arm is attached to a mounting point on the head.

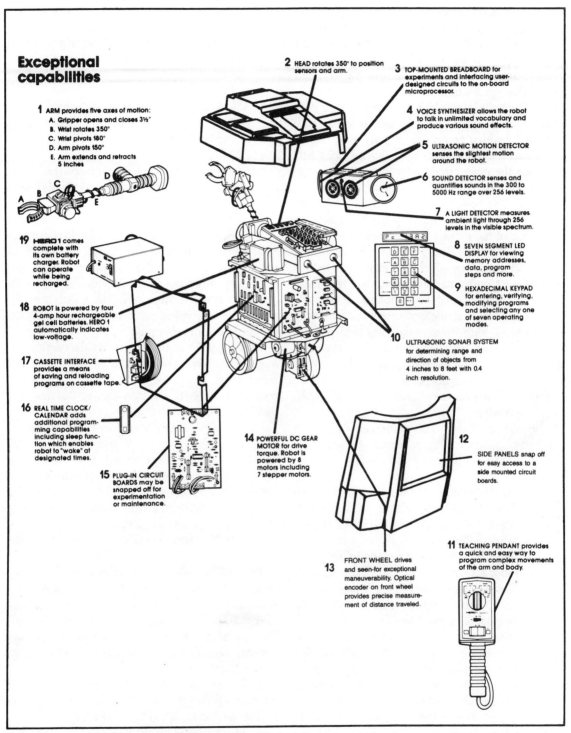

Exceptional capabilities

1 ARM provides five axes of motion:
   A. Gripper opens and closes 3½"
   B. Wrist rotates 350°
   C. Wrist pivots 180°
   D. Arm pivots 150°
   E. Arm extends and retracts 5 inches

2 HEAD rotates 350° to position sensors and arm.

3 TOP-MOUNTED BREADBOARD for experiments and interfacing user-designed circuits to the on-board microprocessor.

4 VOICE SYNTHESIZER allows the robot to talk in unlimited vocabulary and produce various sound effects.

5 ULTRASONIC MOTION DETECTOR senses the slightest motion around the robot.

6 SOUND DETECTOR senses and quantifies sounds in the 300 to 5000 Hz range over 256 levels.

7 A LIGHT DETECTOR measures ambient light through 256 levels in the visible spectrum.

8 SEVEN SEGMENT LED DISPLAY for viewing memory addresses, data, program steps and more.

9 HEXADECIMAL KEYPAD for entering, verifying, modifying programs and selecting any one of seven operating modes.

10 ULTRASONIC SONAR SYSTEM for determining range and direction of objects from 4 inches to 8 feet with 0.4 inch resolution.

19 HERO 1 comes complete with its own battery charger. Robot can operate while being recharged.

18 ROBOT is powered by four 4-amp hour rechargeable gel cell batteries. HERO 1 automatically indicates low-voltage.

17 CASSETTE INTERFACE provides a means of saving and reloading programs on cassette tape.

16 REAL TIME CLOCK/CALENDAR adds additional programming capabilities including sleep function which enables robot to "wake" at designated times.

15 PLUG-IN CIRCUIT BOARDS may be snapped off for experimentation or maintenance.

14 POWERFUL DC GEAR MOTOR for drive torque. Robot is powered by 8 motors including 7 stepper motors.

12 SIDE PANELS snap off for easy access to a side mounted circuit boards.

13 FRONT WHEEL drives and seen-for exceptional maneuverability. Optical encoder on front wheel provides precise measurement of distance traveled.

11 TEACHING PENDANT provides a quick and easy way to program complex movements of the arm and body.

Anatomy of the HERO I (courtesy Heath/Zenith).

Arm: The arm rotates 350 degrees in a horizontal plane with the robot's head. It rises and lowers 150 degrees in the vertical plane and extends and retracts up to 5 inches. The gripper pivots 90 degrees above and below the axis of the arm and rotates through 350 degrees. It opens to a maximum of 6 inches and closes with a force of 5 ounces. The arm can lift a maximum of 16 ounces in the fully retracted position and 8 ounces fully extended.

Torso: The robot's torso provides mounting surfaces for the main circuit boards and houses the main wiring harness and computer.

Base: The base houses the mechanical components of the main drive and steering mechanisms and three of the robot's four batteries.

Computer: The Heath HERO I Robot uses a 6808 microprocessor which interfaces with the robot's sensors, on-board real-time clock, experimental circuit board, and drivemotors. It can be programmed through the keyboard mounted on the robot's head, with its external handheld remote-control teaching pendant, or with programming previously stored on a conventional audio cassette tape recorder. It is capable of storing programs with over 1000 individual steps. An optional RS-232C port is available.

Operating Modes. The computer has seven operating modes, each accessed through the keyboard. Operating modes include:

- ☐ The Executive Mode—the mode from which all other models are entered and to which the computer returns when leaving other modes.
- ☐ The Utility Mode—the mode in which the robot's various utility functions are accessed, such as initializing (robot orientation), arm honing (arm orientation), downloading and uploading programming from cassette tape storage, and setting time and date on the robot's real-time clock.
- ☐ Program Mode—one of the two modes in which the computer is programmed through the keyboard.
- ☐ Repeat Mode—the mode in which stored programs can be accessed and repeated.

The computer can also be programmed through the keyboard in this mode.

- ☐ Manual Mode—the mode in which the robot can be controlled, but not programmed, with the external, remote-control teaching pendant.
- ☐ Learn Mode—the mode in which inputs from the teaching pendant are stored in memory and can be repeated in the Repeat Mode.
- ☐ Sleep Mode—a power conservation mode which allows the user to program HERO I to perform preset tasks on an intermittent schedule. When the robot is in the sleep mode, all systems except the clock are shut down. At preset times, the robot "wakes up," performs the programmed task, and then returns to sleep mode until the next scheduled time. (HERO I can use this feature when functioning as a security device, for example.)

Display. The LED display lets the operator examine and correct programs in the robot, examine information from the sensors, and verify the operating mode. It will also display the time and date upon command.

Batteries. Four 4-amp hour rechargeable gel cells, connected in two electrically independent 12-volt systems (logic system and drive system). The batteries are protected against total discharge by an automatic low-voltage sensor, which shuts down most robot functions (subject to user override) if either system falls to 10 volts.

Table 2-2 shows specifications for HERO I.

The Heathkit Robotics Education Course
Designed to be used with the HERO I robot is a separate, complete robotics education course, a two-volume, learning program individualized for self-learning at home or school. Heathkit's Douglas Bonham says the course can be purchased and pursued without the robot.

"However," says Bonham, "using the robot with the course brings the learning process to life. HERO I is the ideal learning aid for robotics. By letting the individual apply what he or she has just learned,

Table 2-2. HERO I Specifications.

General

Operating Temperature	0C to +40C (32F to 104F) ambient.
Weight	39 lbs. (17.6 kg), with accessories.
Dimensions	20″ high × 18″ diameter (max). (50.8 × 45.7 cm).
Minimum Turning Radius	12″ (30.5 cm).

Charger

Power Requirements	120/240 volts AC, 50/60 Hz, 60 watts (max).
Output Voltage	27 volts (max.) unregulated.
Output Current	1.9 amps (max.) into fully discharged batteries.

you get the kind of reinforcement that makes 'learning-by-doing' the most effective education method ever devised."

The program, which will sell for $99.95, will be published in a two-volume, individual learning program. Bonham says the robotics course is the most extensive program ever developed by Heathkit/Zenith Educational Systems. "It's approximately 1,200 pages long and fills two three-ring self-study texts," he says. "The program consists of eleven separate units, each fully illustrated with charges and diagrams." The course covers the following eleven subject areas:

☐ Robot Fundamentals
☐ Alternating Current and Fluidic Power
☐ Direct Current Power and Positioning
☐ Microprocessor Fundamentals
☐ Robot Programming
☐ The Heath Robot Microprocessor
☐ Data Acquisition (Sensors)
☐ Data Handling and Conversion
☐ Voice Synthesis
☐ Interfacing
☐ Industrial Robots at Work

Programmed self-study texts allow the student to progress at his own pace. The materials are designed to guide the student, step by step, until important concepts are mastered. To help the student understand what has been studied, at the end of each unit are self-test reviews. The course also includes hands-on experiments utilizing the HERO

I robot and a complete final examination.

Heathkit estimates that it will take a person approximately 100 to 125 hours of independent study in 2-hour increments to complete the course and cautions that it is not designed for the technological novice. Therefore, it is best to come to the course already familiar with the basic concepts of electronics. A grounding in mechanics could also be helpful.

HERO JR.

Described as the "new generation robot," HERO JR. is being billed as a new family member by Heathkit, which also claims he is the first affordable, fully preprogrammed personal robot.

HERO JR., unlike most robots, requires no programming skills to operate. A home, or "companion," robot, HERO JR. has a number of unique preprogrammed activities which shape his "personality." He roams, explores, sings songs, recites poetry, and speaks English and his native "Roblish," a robot's version of English. He will wake you up in the morning, guard your home with a coded security system, and even play games.

"HERO JR. is a fun robot with a built-in personality. He's sentimental, sophisticated and at times a real ham," says Wayne Wilson, Product Manager, General Consumer Products.

HERO JR. wakes up his owners with a personalized alarm and can sense whether or not they awaken. Friend and companion that he is, HERO JR. permits a 10-minute snooze.

HERO JR. is programmed to greet his compa-

The HERO JR. home robot greets his family with "I am HERO JR., your personal robot. I am your friend, companion, and security guard." Heath Company's HERO JR. requires no programming skills to operate and has a number of unique preprogrammed activities which shape his personality (courtesy Heath/Zenith).

nions with such phrases as "I am HERO JR., your personal robot," and "I am your friend, companion, and security guard." In addition, an internal 100-year clock permits HERO JR. to remind his owner of the day of the week, date, and time . . . and it even corrects itself twice a year for Daylight Saving Time.

Once familiar with all facets of HERO JR.'s personality, the owner may advance to the level of "Robot Wizard," which permits HERO JR. to identify

the owner by name and much more.

HERO JR. even uses his senses to seek out his owners while moving about. The robot's ability to locate humans can be enhanced with the optional infrared motion detector (standard on some models).

HERO JR. guards your home against intruders when the security mode is selected by issuing a verbal warning and requesting a password. In addition, the Heath GDA-2800-3 security transmitter permits HERO JR. to activate the Heath burglar alarm when an intruder is detected. Included with this accessory are two window stickers that read "Warning; This area is protected by a security robot."

HERO JR. enjoys playing games, including "Cowboys and Robots," "Let's Count," and "Tickle Robot." Additional cartridges that expand his operation are also available. These cartridges teach, play games, and add to his repertoire of phrases and songs.

Using HERO JR.'s optional wireless remote control, his owner can drive the robot from place to place. Otherwise, HERO JR.'s normal mode allows him to explore at random, avoid obstacles, and seek out humans.

HERO JR. is preprogrammed to sing *Daisy* and *America the Beautiful.* The accessory cartridges enable him to sing other songs too. A built-in demo program allows HERO JR. to show off his many talents in a "Robot Variety Show" in which he demonstrates the numerous personable tasks he can perform.

A battery accessory doubles HERO JR.'s operational time, which is normally 4 to 6 hours. The batteries recharge overnight from a plug-in wall charger.

HERO JR. is priced at about $1,000 fully assembled. A kit is available for about $600.

A Description of HERO JR. HERO JR. is a fully preprogrammed personal robot which comes with 32K of built-in robot routines. His accessories are exceptional, such as the infrared motion detector and a wireless remote control for manual operation. Let's take a close look at HERO JR. and learn all about the machine "person."

Exterior: HERO JR. is 19 inches tall, weighs 21.4 pounds and resembles Heath's first robot, HERO 1, which is designed to teach robotics and industrial electronics.

HERO JR.'s three wheels, including a single articulated rear drive wheel, enable him to move about and avoid obstacles. He can carry up to 10 pounds on a 94-cubic-inch compartment built in to the top of his head. A cartridge is located in the back of this tray.

His head is also equipped with a 17-key keypad which permits the owner to modify HERO JR.'s per-

Heath's able robot, HERO JR., is a home/personal robot. Using preprogrammed computer memory, HERO JR. is ready to run a wide variety of robot routines at the push of a button (courtesy Heath/Zenith).

sonality or initiate a special task. Eight data LEDs flash in time with his speech or to signal something special.

A window for optional infrared motion detection, an ultrasonic Polaroid sonar transceiver, and a light sensor compose HERO JR.'s face. Other transducers provide synthesized speech output and sound sensing.

Connectors on the back of HERO JR.'s head include a charger jack, an on/off slide switch, sleep switch, and an RS-232C computer interface. HERO JR. can be recharged overnight from a plug-in wall charger which comes with the robot. HERO JR. will operate from 4 to 6 hours between charges under normal operating conditions.

Sensors: For sound sensing, HERO JR. uses a 256-bit resolution sound sensor with adjustable range and a 200 to 5000 Hz bandwidth. He uses a

256-bit resolution light sensor with adjustable range and a 25-degree reception angle. His ultrasonic sonar is designed to accurately measure distance from 4 inches to about 25 feet.

HERO JR.'s standard motion detection sensor also uses his ultrasonic sonar. The optional six-field infrared sensor, however, provides superior heat/motion detection capability and improves his ability to seek out humans.

HERO JR.'s speech synthesis equipment includes a Votrax SC-01, which 4 pitch levels and 64 phonemes, which permits HERO JR. to say just about anything.

HERO JR.'s time apparatus consists of a CMOS processor, which includes a clock with a 100-year calendar and automatic correction for Daylight Saving Time. An RS-232C computer interface also allows HERO JR. to accept an assembler or load and dump from memory. A "HERO JR. Basic" cartridge permits programming through the RS-232C interface.

Accessories: An infrared motion detector helps HERO JR. locate humans and improves operation of the security mode and some of his games. Wireless remote control enables him to be operated manually.

The Heath GDA-2800-3 RF security transmitter accessory permits HERO JR. to activate the Heath GD-2800 burglar alarm system, should the robot detect an intruder who is unable to signal the proper password.

Cartridges can be added to expand HERO JR.'s repertoire of routines, songs, games, and phrases. An extra battery accessory doubles his operating time.

Software: HERO JR.'s software operation consists of a preprogrammed or built-in personality, which requires no user input. The user can shape HERO JR.'s personality by increasing or decreasing the priorities of each of the six personality traits, or he can select any individual task or demonstration mode separately.

The software program includes four special task commands. *SETUP* is for changing HERO JR.'s personality. *GUARD* commands him to protect a specific area or to act as a security device while moving about. *ALARM* commands him to wake up his owner at a designated time. *PLAN* permits you to set HERO JR. for a future activity, such as reminding you of a birthday or anniversary.

The software performs true multi-tasking, which enables HERO JR. to move and speak at the same time.

Electronic Specifications: HERO JR. uses a Motorola 6808 microprocessor. He has 32K of monitor ROM, 8K RAM, and an on-board provision for up to 16K additional RAM or ROM for future expansion, plus a provision for 4K and 8K plug-in ROM cartridges.

HERO JR.'s three circuit boards include a microprocessor, power supply/sense, and keyboard. He comes with two Motorola 6821 parallel interface adapters and a Motorola 146818 CMOS clock.

He uses a 180-degree rotation, stepper type motor for steering, and a 12vdc motor for drive. One idler wheel features optical chopper feedback to detect distance traveled. His alphanumeric keypad features clearly marked function keys including Sing, Play, Poet, Gab, Alarm, Guard, Help, Demo, Plan, Set Up, and Enter.

Power Supply: HERO JR. uses two 6-volt, rechargeable lead-acid gelled-electrolyte batteries. Two optional cells double his operating time. A 12-volt wall charger plug provides full recharge overnight while HERO JR. is in the sleep mode.

Questions and Answers About HERO JR. To get a feel for what the Heathkit Company had in mind when they developed HERO JR., here are questions posed to the company about the robot and their answers. If, after reading the narrative and description just given, you still have questions about HERO JR., these questions and answers will fill in the knowledge gap.

Keep in mind that, quite naturally, Heathkit contends that its robot is "the first," "the best," and "the only" robot to perform certain tasks or to possess certain characteristics. Some other robot manufacturers may, however, arguably make these same claims for their own products. The personal robot field is young and growing fast, and new developments are coming at a breakneck pace; so it is sometimes difficult to credit one company or the other for technological achievement.

☐ Who is HERO JR. and what does he do?
HERO JR. is a friendly, fully preprogrammed robot intended for home and
family use. He is primarily designed for
entertainment. He can sing songs, play
games, tell nursery rhymes and recite
poems. He is "friendly" because he will ac-
tively use his sensors to explore his surroundings. Without supervision or help, he
will seek to remain near his human companions. HERO JR. has another important
use. With his accurate sensors, he is a no-
nonsense home security guard. Heath Company includes security notice stickers with
the optional infrared motion detector (standard on some models).

☐ What is meant by HERO JR.'s "personality?"
Every attempt has been made to make
HERO JR. take on human characteristics.
The result is that HERO JR. is the first robot with a dynamic personality. HERO JR.'s
personality consists of six "traits." He sings
songs, speaks English phrases, plays
games, explores his environment, seeks out
humans, gabs in his own language,
"Roblish," and tells nursery rhymes and
poems. HERO JR. is the first completely
preprogrammed robot. As soon as he is
turned on, he will randomly select one or
more of these traits and act out his personality. What HERO JR. will do next will be
a complete surprise—another humanlike
characteristic. No other robot can do this.

☐ Can the owner create a unique personality
for HERO JR.?
Yes. There are thousands of unique personalities that can be created by merely
pushing a button. Called the "setup key,"
this switch produces a voice prompt, or
menu, that presents each personality trait
and asks the owner to set a level of activity
from zero to nine for each trait. For example, to prepare for a party, the owner can
alter HERO JR.'s personality so he will do

very little exploring but a lot of singing and
playing games. Plug-in cartridges give
HERO JR. even more capability.

☐ What kinds of cartridges are available?
There are quite a few cartridges for HERO
JR, for example Math master, Songs,
Phrases and Rhymes #1, Animals, Blackjack, Tic Tac Toe, and Riddle Robot,
Tongue Twister. For special occasions he
will sing "Happy Birthday," "Auld Lang
Syne." There is HERO JR. BASIC: allows
the owner to program additional personality through a home computer. Several outside software development houses are
currently preparing additional cartridges.

☐ If HERO JR. is preprogrammed, why does
he have a 17-button keypad?
The keypad permits the owner to change
HERO JR.'s personality. The keypad also
allows the owner to get HERO JR.'s attention and request a specific task. Also, the
"HELP" key activates an audio menu that
will help the owner set activities.

☐ What kind of vocabulary does HERO JR.
have?
He has an unlimited vocabulary. His voice
synthesizer can closely duplicate all English
sounds. He can even speak phrases of most
foreign languages.

☐ How is the RS-232C computer interface
used?
Heath/Zenith introduced a HERO BASIC
cartridge that allows the owner to program
HERO JR. through a terminal or home
computer. This version of BASIC contains
special enhancements for speech, movement, and other traits.

☐ Why doesn't HERO JR. have an arm?
HERO JR. is a personal robot, priced for
home use. There simply is no robot arm
anywhere on the market that is practical or
functional for the home. To get any use
from an arm, a robot requires an extremely expensive navigation system. This would
make HERO JR. too expensive for the
consumer.

☐ How much does HERO JR, cost?
About $1,000.

☐ Is there a market for home robots?
Yes. Heath Company is currently the leader in the innovative stages of the personal robotics industry. The company expects this market to blossom by the end of the decade.

☐ Who buys HERO JR.?
General consumers, especially electronics and computer game enthusiasts and the type of consumer who buys a satellite receiver, a VCR with Hi-Fi or compact disc audio. HERO JR. also appeals to parents who want a robot to introduce their children to computers and robots.

☐ Where is HERO JR. sold?
HERO JR. is sold by upscale department stores and retailers that sell computer games and software. He is also sold in some of the new robotics stores that are now appearing. Heath/Zenith is currently selecting regional and national distributors.

☐ Is HERO JR. a replacement for Heath Company's first robot, HERO 1?
No. HERO 1, the world's largest-selling robot, is specifically oriented as a training robot for schools, industry, and serious hobbyists. HERO JR. was specifically designed for the home and personal robot market.

☐ Why did Heath/Zenith design HERO JR.?
Through Heath's customer experience with HERO 1, the company leraned there was high demand for a lower-cost personal robot. Customers wanted a robot that did not require special technical skills or programming knowledge to operate. HERO JR. does not need any programming.

☐ Are there educational applications for HERO JR.?
Heath Company foresees use for HERO JR. as a learning motivator in elementary schools. He will serve as an exciting introduction to computers and robotics.

☐ Is Heath Company planning a HERO JR. II?

Heath Company is definitely in the personal robot business to stay. It can be safely assumed that the company will be introducing more robotic products that will fill other special niches in the market.

☐ How many HERO JR.'s does Heath/Zenith expect to sell?
A lot. HERO I was targeted for the relatively narrow education robotic market. Yet HERO I has become the world's most successful robot. HERO JR. is designed for a much broader market, and Heath Company expects to sell considerably more robots in this market.

☐ Whose idea was HERO JR.?
HERO JR. was a collective idea and a logical extension of a robot into the general consumer market.

☐ Can you please list for the readers of *The Personal Robot Book* the basic attributes of the robot?

PHYSICAL DESCRIPTION
—19 inches high, weighs 21 1/2 pounds.
—3 wheels. One rear wheel steers and drives.
—2 rechargeable batteries. Recharger included.
—Built-in tray carries up to 10 pounds.

SENSORS
—Sound: 200 to 5000 Hz bandwidth.
—Light: 30-degree reception angle.
—Ultrasonic Sonar: judges precise distances from 4 inches to about 25 feet.
—Motion Detection, Standard: uses ultrasonic sonar.
—Motion Detection, Optional: uses infrared motion sensor.
—Time: uses CMOS clock with 100-year calendar. Compensates for Daylight Saving Time.

PERSONALITY TRAITS/CAPABILITIES
These traits are available standard, when HERO JR. comes "out of the box." No programming is needed. Additional games, songs, etc., are available through optional cartridges.
—Sings: *America The Beautiful* and *Daisy*

—Speaks 18 English phrases.

—Plays games: Cowboys and Robots, Let's Count Hand Claps, Tickle Robot

—Explores: Moves about, using sensors to avoid obstacles. Like a pet, will try to stay in company of humans.

—Gabs: Speaks random voice phonemes that sound like English.

—Poet: tells nursery rhymes from memory.

—Wake up alarm: Awakens human at selected time, listens to be sure he is awake. Permits two 10-minute snoozes.

—Self Demo: Demonstrates sensors and speech to owner's friends. Shows HERO JR's capabilities.

—Security Guard: Will guard specific area or room using sensitive motion detector. If intrusion is detected, shouts warning, demands password, and appropriate action.

How to Buy the Heathkit Robots

HERO JR., Heath Company's companion robot, can be purchased fully assembled at upscale department stores and other retail outlets that sell computer games and software. HERO I is also for sale at many retail outlets. Either the HERO JR. or the HERO I can also be purchased in kit form at Heathkit Electronic Centers, located in most major metropolitan areas. Nationwide, there are about 53 such centers.

In addition, the HERO robots can be ordered by mail through a Heathkit catalog or by phone directly from Heath Company. Heathkit catalogs are available free from Heath Company.

Schools, industries, and government organizations may order a fully assembled HERO JR. through Heath Company's educational representatives located nationwide. For more information, write or call Heathkit/Zenith Educational Systems Division.

Two RB5X personal robots, presumably carrying on an interesting conversation (courtesy RB Robots, Inc.).

RB ROBOT

In September, 1982, the RB Robot Corporation of Golden, Colorado announced the introduction of the lovable and functional RB5X, advertised as the "Intelligent Robot." This momentous event was noteworthy because RB5X was the world's first mass-produced personal robot. Since then, the tiny, 23-inch-tall robot has proven to be one of the most reliable and well-made machines on the market. As a result of this reliability, RB Robot Corporation has sold thousands of RB5Xs to customers in the United States. Distributors in Germany, Japan, and the Far East have also sold many units in their home countries. (In West Germany, RB5X is modified slightly and has been redesignated as "Toby.")

RB5X resembles a high-tech garbage can, with his cylindrical body and a clear plastic domed top. As he scoots along, propelled by two motor wheel

assemblies, four randomly flashing LEDs add a touch of brilliance.

RB5X has an optional arm and hand, or *gripper*, which he cleverly conceals in his body. Like HERO I, RB5X isn't much in the muscles and brawn department. A little less than 1 pound—maybe a can of Coke, your slippers, or the morning paper—is about all the small fellow can lift. The arm has five axes of movement.

A nice feature of the robot is his built-in sonar and bumper switches that help him find his way around a room or area by trial and error. As RB5X moves to and fro groping his way, his memory records the best—that is, the correct—path.

RB5X is a user-friendly robot that has impressed many observers, including robotics and computer experts. The little robot was a big hit at the 1984 Winter Consumer Electronics Show in Las

This dapper RB5X robot shows he knows how to romance the ladies. . .with a bouquet of flowers, naturally (courtesy Marilyn Chartrand).

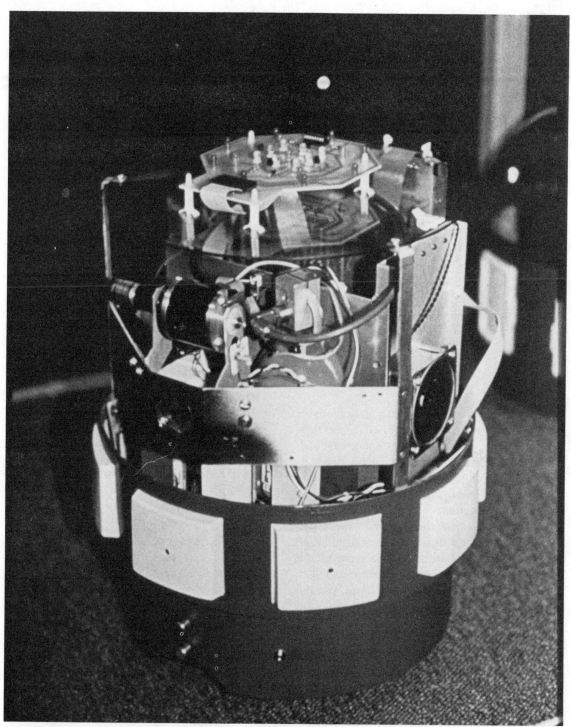

RB5X shown with optional fire detector/extinguisher package. Many attachments and add-ons are in the works for this robot (courtesy Marilyn Chartrand).

Vegas. At that show, the RB Robot people presented a most unusual display. The centerpiece of the display was a "sculpture" of six, multilevel pedestals, each presenting a robot performing a different task. One pedestal featured an RB5X singing, *Daisy*, the song made famous by HAL, the computer in the film *1001: A Space Odyssey*. Another pedestal held an RB5X that greeted show visitors with a tip of its hat. Another RB5X entertained with the robot version of a carnival barker ("Come one, come all!") In addition, there was a plant-watering robot, and two RB5Xs who passed a boquet of flowers back and forth.

Visitors to the RB Robot booth were able to gain hands-on experience with the RB5X robotic arm and with a software module called "Pattern Programmer" that allows users to program the RB5X to move in any pattern they design by pressing its bumpers. Visitors also saw an RB5X equipped with heat sensors and a fire extinguisher that will soon enable it to detect a flame, seek it out, and douse it with Halon.

About RB5X

RB5X comes with his own programmable microprocessor and also has an RS-232C interface to permit link-up with a personal computer for added memory. Because the robot is gaining popularity and becoming a mobile fixture in more households, independent software firms are bringing on-the-shelf software packages to the market which continue to make the robot more useful and fun. In addition, the company that makes RB5X offers a number of software packages and useful add-ons, including voice synthesis and voice recognition.

Looking through RB5X's clear head, you can see the slot for his National Semiconductor on-board microprocessor, INS 8073. On-board capacity is 8K and can be boosted to 16K. There are also five other slots: one to hold extended memory and four additional slots to permit the connection of other boards.

With the robot you get software demonstration programs on disk, which enable you to give the robot a good workout and acquaint you with RB5X's overall capabilities.

Among the optional items offered is a Robot Control Language (RCL)™ package that uses Savvy™, a system that lets you program RB5X with common English words and phrases. RCL is available for Apple II and *II*e computers. Contact RB Robot Corporation for availability on other personal computers as well.

The voice synthesis and recognition features allow interactive conversation between a person and the robot. Using the 64 phonemes standard on a speech synthesis microchip, the user can create a varied vocabulary in any language. Also, you can command the robot to take actions and hear RB5X respond by voice. New communications packages offered by independent firms, such as Arctec, go one step further, permitting RB5X to communicate with other robots, including the HERO models.

Joe Bosworth, president of RB Robot Corporation, feels his robot is one of the most useful on the market, recommending it for home or school use, for play and education, and for experimentation. He says that one way to view the robot at this stage is to "think of it as an infant." Like a human baby, remarks Bosworth, a baby robot "evolves and changes as it gains more experience and knowledge." RB5X is growing, adds Bosworth, is becoming an adolescent, and within the next few years, may become a full-fledged adult. All of this is to say that RB5X is getting more sophisticated and improved with the passage of time.

Improvements to Come

Planned improvements to RB5X include several attachments, including a fire detector/extinguisher and a vacuum cleaner. The company has built successful prototypes of each of these devices. The fire detector/extinguisher has a "nose" that smells smoke. Sensing a fire, it then targets the location and puts the fire out with a fire extinguisher it handily carries on-board for just such a dangerous occasion.

The RB5X vacuum attachment includes programming that will allow the owner to preprogram the robot with a knowledge of the room or rooms that need vacuuming. Then after the master of the house has left for work, the robot turns on automatically and proceeds to complete his appointed cleaning chore. This attachment is to have its own motor and batteries. Its replacement bags are standard so they can be purchased at the neighborhood supermarket or variety store.

Two other useful add-on items to come are a communications software package to allow RB5X to "talk" to a remote computer, or even another robot, and a trailer that can be hooked up to the robot. With the wheeled trailer, RB5X can pick up and deliver mail, haul small loads (such as laundry), and perhaps, take the children on a joy ride. Another add-on item available soon is a new programming package for the Commodore 64 computer that features direct, text-to-speech conversion and EPROM programming capabilities.

RB5X Features and Options

Following are the basic features you will get with your purchase of a RB5X. A description of seven options that are currently available is listed afterwards.

On-Board Microprocessor. Unlike remote-control robots, the RB5X has an INS 8073 microprocessor built in, making it completely programmable and independent. Owners use a computer keyboard and screen to write programs for the robot, and then download them to the RB5X's microprocessor.

Self-Learning Software. The RB5X comes complete with Alpha and Beta levels of self-learning software, which enable the robot to learn from its experiences. Developed by robotics author David Heiserman, this software allows RB5X to progress from simple random responses to an ability to generalize about the features of its environment, storing this data in its on-board memory.

Space for Additional Electronics. One of the RB5X's special features is an interior card cage that allows for the addition of up to four circuit cards. This flexible design enables users to enhance their RB5Xs with special hardware and to make each RB5X virtually one-of-a-kind.

Tiny BASIC. The RB5X's "native tongue" is Tiny BASIC, a subset of the BASIC language that

is both high-level and easy to use.

Sonar Sensor. The RB5X comes equipped with the Polaroid Rangefinder™ sonar sensor, which allows the robot to detect objects in its path as it moves. The sonar detection range can be set from 10 inches to 35 feet from the robot and is programmable.

Tactile Sensors. Eight tactile sensors, or bumpers, ring the skirt of the robot, allowing it to sense when RB5X makes contact with another object. Like its sonar, this feature allows the robot to navigate its way through its environment.

Autonomous Battery Charging. A special circuit in the RB5X enables it to recognize when its battery charge is low and to begin using special software that helps it find its charger. The RB5X uses its photodiode system to locate its battery charger, moves against the charger nest, recharges itself, and then automatically resumes its activities.

Battery Shutdown Circuit. RB5X comes with a special circuit that shuts the robot down if its batteries drain close to the point where they cannot be recharged. The robot cannot be switched on again until it has been recharged. A charge indicator on the interface panel shows the battery charge level.

Software Module Socket and Switch. RB5X's interface panel contains a socket for preprogrammed software modules. A switch allows owners to set the robot for modules of either 2K or 4K.

Utility Software Module. A standard feature of RB5X is a 2K utility software module that contains a self-diagnostic routine (which automatically checks the robot's batteries and motors), as well as several of the robot's standard software routines.

Dual RS-232 Interfaces. RB5X has two RS-232 ports for handling communications with computers and with other, future options for the robot.

Options Cutouts. RB5X's upper body contains a series of cutouts, covered by removable plastic caps, that accommodate hardware attachments.

Programmable Lights and Horn. RB5X's

pulsating lights not only enhance the appearance of the robot, but can be programmed to correspond to whatever mechanical or electronic events the user designates. This feature, along with RB5X's standard horn, can alert the owner to special circumstances or can be used simply for extra interest.

Extended Memory. The RB5X extended memory option is a circuit board that plugs into the interior card cage, adding 16K of RAM to the robot's standard 8K.

The RB Arm. A robotic arm, which extends from a resting position completely inside the robot's body, turns the RB5X into a messenger able to carry objects weighing up to 16 ounces. The arm can be maneuvered under direct program guidance, using a controller to manually guide it, or using a controller and an arm training software module.

Voice/Sound Synthesis. The RB5X voice/sound synthesis option enables owners to program their robots to speak and to make a variety of sounds. This package contains a speaker and a printed circuit board with pitch and volume control.

Voice Recognition. For owners of Apple II+ computers, there is a voice recognition option available that enables the RB5X to respond to spoken commands through the Apple.

Software Modules. Owners may either program RB5X using a computer, or they may purchase preprogrammed software modules. These modules enable RB5X to do specific tasks as soon as the user switches the robot on.

Robot Control Language with Savvy (RCL). Robot Control Language with Savvy (RCL, for short) is a software development language that enables RB5X users to program their robots using common English words and phrases. Currently available for use in conjunction with Apple II+ and Apple *IIe* computers, RCL will soon be available for the IBM PC.

Power Pack. This option allows an owner to extend RB5X's charge life and, thus, the run time of the robot. A 10-amp hour battery that attaches to the robot's existing battery, the Power Pack, can keep the base RB5X running for up to 10 hours or the RB arm alone for up to 2 hours.

The Optional Software Packages

To give you an idea of the educational uses RB5X can facilitate and the fun the owner can have, it is necessary to take a look at a few of the optional software packages available from the RB Robot Corporation. So, let's briefly cover some of these packages, the first being the "RB5X Terrapin Logo Translator," a software system that allows RB5X to execute turtle graphics procedures, making the robot an education tool for demonstrating the physical manifestations of Logo programming. Suggested retail is $34.95.

"Bumper Music" is a limited but interesting program that lets you play musical notes by pressing the bumpers which ring the robot's outer shell. Each of the eight bumpers plays a different note, and you can create music by working the bumpers in different combinations.

Two RB5X software modules, "Hop to It!" and "Math Whiz," are application programs on erasable, programmable read-only memory (EPROM) chips that plug into a socket on the robot's interface panel.

"Hop to It!" is an engaging, educational game that allows RB5X to use its sonar sensor to challenge players to accurately judge distances in feet and inches. RB5X issues a verbal challenge, records and calculates players' scores, and announces the winner. Suggested retail is $24.95.

"Math Whiz" is a math quiz in a game format that may be played by up to eight people at one time. RB5X uses its random number generator to compose a math problem involving elementary addition, subtraction, multiplication, or division for each player in turn. The robot checks players' answers for errors, corrects or congratulates, and calculates players' scores. This program's lights, sounds, and robot motions motivate children to learn important math facts with RB5X. Suggested retail is $24.95.

"Intruder Alarm/Daisy Daisy" is a package for the RB5X equipped with optional voice/sound synthesis capability. With the Intruder Alarm, a Polaroid sonar sensing device picks up movement within its range, and the robot then sounds the alarm. *Daisy Daisy* is a bit more joyful, allowing RB5X to sing the song made famous in the movie, *2001: A Space Odyssey.*

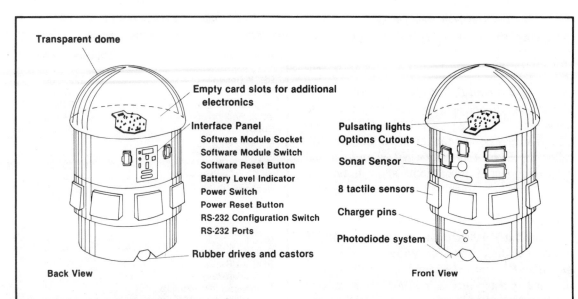

Transparent dome

Empty card slots for additional electronics

Interface Panel
Software Module Socket
Software Module Switch
Software Reset Button
Battery Level Indicator
Power Switch
Power Reset Button
RS-232 Configuration Switch
RS-232 Ports

Rubber drives and castors

Back View

Pulsating lights
Options Cutouts

Sonar Sensor

8 tactile sensors

Charger pins

Photodiode system

Front View

PREREQUISITES FOR USE

Can be programmed using any brand or model of computer, equipped with selected communications software, or operated using preprogrammed software modules designed specifically for the RB5X.

RB5X SPECIFICATION.

Dimensions:	13" in diameter x 23" high	Programming Languages:	Tiny BASIC™ Robot Control Language™ with Savvy® optional
Weight:	24 lbs.		
Speed:	Approximately 4" per second		
Sensor Inputs:	8 bumper panels around perimeter 1 photodiode under carriage 1 sonar transducer	Power Supply:	6V 5 amp-hr. sealed lead acid battery for electronics 6V 10 amp-hour sealed lead acid battery for motors
Controlled Devices:	2 motors 5 programmable light-emitting diodes (LEDs) 4 randomly flashing LEDs 1 horn sonar		6V 10 amp-hour sealed lead acid battery Power Pack optional Battery self life is a minimum of 1000 days.
Microprocessor:	National Semiconductor INS8073	Recharging Nest:	7.5V 750ma; operates on 110V AC 60 Hz
I/O System:	National Semiconductor INS8255 6 eight-bit bytes of I/O to 4 edge connectors including + 5 volts and ground. Edge connectors: standard 2 x 22 pins on .156" centers.	Construction:	Aluminum chassis, polycarbonate dome
		Wheels:	Two 4" diameter synthetic rubber drives and Two 2" diameter castors
Memory:	8K random-access memory (RAM) standard 16K optional	Standard Software:	Alpha and Beta self-learning programs, sonar software, charger finder routine, and utility software module including diagnostics. Preprogrammed software modules available as options.
Interface:	Serial RS-232		
Data Transfer Rate:	Adjustable to 110, 300, 1200, or 4800 baud		

Savvy is a registered trademark of Excalibur Technologies Corp.

RB5X: The basics (courtesy RB Robot Corporation).

RB Robot Price List

Table 2-3 is RB5X robot price list effective August 1, 1984.

RB's Robot Appreciation Kit

If you're new to personal robotics and you want to know more about the personal robot field, RB has a kit that may be just the thing. Actually, the kit, called the *Robot Appreciation Kit,* is an item that even experienced robot hobbyists will find interesting. It's designed to answer your questions about home robots in general and RB5X in particular.

Priced at $19.95 and available from local RB5X retailers or direct from RB Robot, the *Robot Appreciation Kit* contains an overview of the field of personal robots; article reprints from current periodicals; product literature on the RB5X; a copy of the July, 1983 issue of *RB Forum,* which discusses Robot Control Language with Savvy; a questionnaire and free *RB Forum* subscription offer; two RB5X bumper stickers; an RB5X poster; a copy of Isaac Asimov's book, *Eight Stories from the Rest of the Robots;* and a copy of the RB5X *Reference Manual,* less the technical appendices. RB Robots refunds the price of the kit to persons who subsequently purchased the RB5X robot.

The Future

RB Robot Corporation merged in October, 1984, with Actronix Corporation, a Dallas, Texas robotics development firm. Reportedly, Actronix has some top-notch management and research and development personnel who will lend their expertise to building the new, combined company into a big success. At the time of the merger, Actronix had already developed prototypes of two personal robots—the *Actron Bear,* an upright device with a 300-pound lifting capacity, and the *Actron Wolf,* a low-profile, mobile security robot. It is possible that one or both of these robots will be available soon. In any case, the research and development knowledge and experience acquired in their development may be transferred to future RB5X models and thus mean enhanced technological advancement.

Table 2-3. RB5X Price List.

HARDWARE	Suggested Retail
RB5X Base Unit	$2295.00
RB Arm	1495.00
Voice/Sound Synthesis	245.00
Voice Recognition	995.00
Robotlab	1395.00
Sensor Kit	245.00
Sensor Kit--Wired	345.00
16K Memory	195.00
Power Pack Add-on	94.00
Recharge Cable Kit	29.95
SOFTWARE	
RCL I	179.00
RCL II	185.00
RCL I w/Savvy	455.00
RCL II w/Savvy	595.00
Curriculum for Robotlab	TBA
Curriculum for K-12	TBA
Terrapin LOGO Translator Disk	34.95
EPROMS 2K	19.95 ea.
Spin the robot	
Pattern Programmer	
Bumper Music	
EPROMS 4K	24.95 ea.
Daisy-Daisy/Intruder	
Voice/Sound Demo	
Carnival Barker	
Nursery Rhymes	
Hop To It	
Math Whiz	
MANUALS	
Robot Appreciation Kit	24.95
Tiny BASIC Users Manual	30.00
User Reference Manual	35.00

How to Buy the RB5X

A growing number of dealers in the United States and Canada are stocking and selling the RB5X and accessories. Also, RB5X is available direct from the manufacturer. For the name of the RB5X dealer nearest you, to order RB5X, or for further information, write or call RB Robot Corporation.

COMRO, INC.

ComRo ToT, the top-of-the-line personal robot produced by Comro, Inc., New York City, had an auspicious beginning. The offspring of ComRo I, a

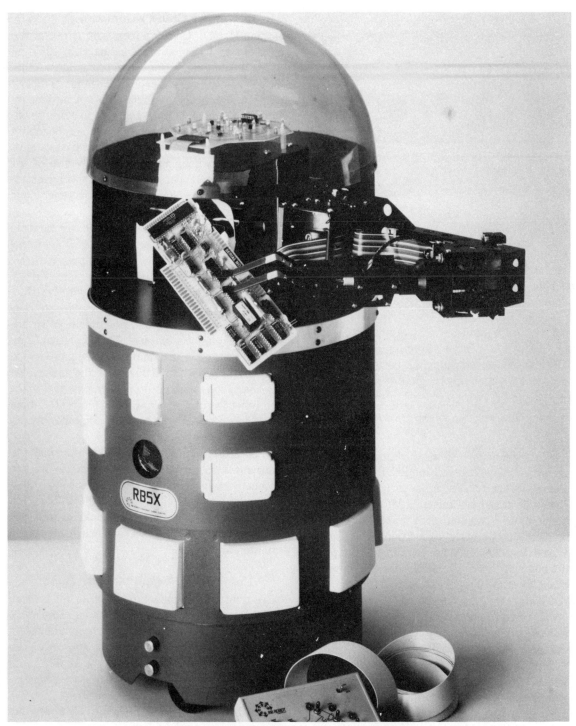

RB5X shows off his manipulator arm and gripper. The box at the bottom is a controller for the robot's actions (courtesy RB Robot Inc.).

The ComRo Company

Jerome Hamlin, ComRo's president, is one of the pioneers of the home robot industry. In 1975, he began building robots for use as promotional actors in TV commercials. Then in 1981, his company made a significant breakthrough when Neiman-Marcus, the famed Dallas department store, decided to feature "his and her" companion robots built by ComRo in its fabulous Christmas catalog. Priced at a whopping $15,000, only two of the robots were sold, but the publicity was a tremendous boon for the home robot field in general and ComRo in particular.

ComRo ToT is the successor to the pair of robots offered in the Neiman-Marcus catalog. Don't worry about that $15,000 price tag, though. With advanced technology and the benefit of increased production, Hamlin has been able to drastically reduce the price of the ComRo ToT, in comparison to its predecessors. What's more, this unique robot has equally magnificent features. You'll find that ComRo ToT costs more than many other robots—for example, HERO I and RB5X—but the maker has packed in some impressive refinements that make the robot worth the asking price (currently about $3,495 for a model with on-board computer control and accessories).

ComRo, Inc. also will build custom robots to your specifications. In addition to ComRo ToT, the company builds and markets another home robot, a smaller, young "fellow" called the Bubble Tot. Reportedly, the company is looking into the possibility of developing robots to help handicapped children and adults.

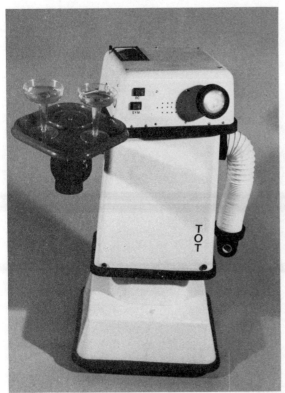

ComRo ToT, a fully programmable robot, can perform a large variety of tasks, including serving drinks (courtesy ComRo, Inc.).

robot the company believes was the first domestic, or home, robot system, ComRo Tot (also called, simply, ToT) was "born" in January, 1983. The first prototype model, designated TMC 3000, was sold to corporate communications giant, Warner Communications. Knowing a good thing, the people at Warner had decided to use ComRo ToT as a promotional gimmick for *The Movie Channel,* a national cable TV system.

Warner conducted a contest, the "GREAT Robot Giveaway" to promote *The Movie Channel's* gala showing of *Star Wars.* The contest was a huge success, drawing over 50,000 entries. ComRo ToT officially presided over the contest and chose the winner. First prize was the robot, with the lucky winner able to take ComRo Tot home. Two hundred other runner-up winners were each given a limited edition T-shirt which had a picture of the robot emblazoned on it.

ComRo ToT: A Multipurpose Robot

According to its manufacturer, ComRo ToT is much more than just a toy or a mere home companion. The robot is also touted as a great promotional tool for companies and organizations and as an expandable education system that can enable youth and adults to learn more about computer control, electronics, and robotics.

ToT as a Mobile Computer. Three-foot tall and weighing in at 50 pounds, ComRo Tot is a fully programmable, mobile computer. Using the popular

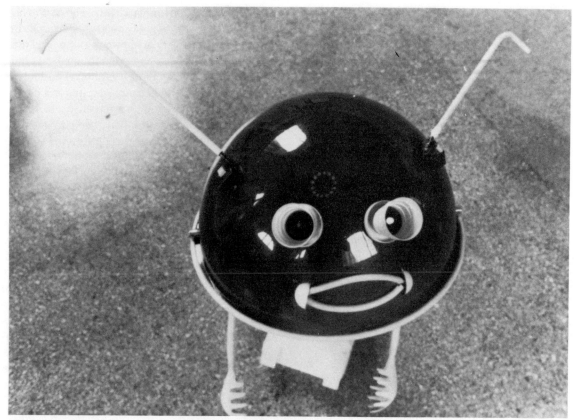

The Bubble ToT personal robot, from ComRo (courtesy ComRo, Inc.).

6502, 8-bit microprocessor and a hexadecimal keypad conveniently located on the top of his head, the robot can be programmed to carry out a number of simple household tasks, such as sweeping the carpet and serving refreshments. With ToT's Votrax "type and talk" speech synthesizer, the facile robot can greet and chat with guests and educate the children. A home security program turns ToT into a lurking, nighttime burglar alert and alarm system. Anyone who passes within range of the robot's ultrasonic ranging system is brought up short when ToT bellows out, "I see you! . . . You are under surveillance."

ComRo ToT comes with many accessories, including a utility bar, carpet sweeper attachment, lifting device, a "payload" bucket, and a tray with a built-in space for spill-proof tumblers. For those who wish to customize their robot, ComRo will build

a home entertainment center in a home consisting of a TV and radio/cassette player.

The ComRo ToT is a physically attractive robot. A handsomely casted case is highlighted by a single lighted electric eye on the front of the robot's rotating head. ToT is the only personal robot that is available in six different colors. So, you can coordinate its color with your home's decor or, perhaps, the color of your eyes.

In addition, ToT is much more sturdy than most other personal robots. His body is made of resin-reinforced fiberglass with steel rod superstructure. ToT is tough enough to be used outdoors on relatively smooth surfaces. His four-wheel drive can negotiate cracks and crevices that stop most other robots in their tracks.

ToT as a Show Robot. ComRo ToT's usefulness as a show, or promotional robot, is enhanced by an optional package (costing about $1,000) that renders the robot operable by remote control, using a handheld radio control pendant. Standing out of sight or behind a curtain or partition, an operator can have ToT perform as a barker of goods, pass out business cards, relate amusing stories and jokes to passers-by, and generally attract scads of attention. Said one male spectator upon observing the huge, bemused crowd at a supermarket where ComRo ToT was performing: "Boy, this thing is more popular with a crowd than a bevy of bathing beauties or even Miss America!"

The ComRo company says that about half the sales of ToT are to promotional users. One prime example is ComRo ToT's work at the New York City Aquarium, where trainers have taught sea lions to respond to commands from the robot. This is done by arm signals from ComRo ToT which the sea lions look for and recognize.

ToT as Educator/Trainer. The robot has much to offer in the way of robotics education. Its four-wheel drive, 360-degree pivot turns, a head that can rotate 300 degrees, and programming capability make ComRo ToT a most useful teacher. Its two arms lift up to 4 pounds and swing in a full circle. Also available as an option is a remote control gripper so nimble it can pick up a coffee cup or mug. (ToT's two regular arms and end-effectors are not of the gripper type.)

Students who work with ComRo ToT will learn the fundamental applications of machine code and the hexadecimal numbering system in the most entertaining way possible. *Machine code*, the most elementary computer language at the heart of almost every computer system, comes alive as students program ToT to walk, talk, stand guard, etc. The 16-key hexadecimal terminal on ToT (and HERO I as well) is somewhat awkward to use at

ComRo ToT is a proven attention-getter. At New York City's Aquarium, he is used by trainers to direct the sea lion act (courtesy ComRo, Inc.).

first, but you get better at it.

Robotics students learn the instruction set for the 6502 8-bit microprocessor. They learn how to access memory locations in ROM and RAM. How to write to interface ports in order to turn on and off external devices is quickly picked up. ToT's real-time software clock, dozens of subroutines, 10 sample user programs—which can be modified or totally recreated—all make ToT a powerful computer/robotics trainer.

The basic ToT user's manual provides a tutorial for programming the robot. In addition, two technical reference guides come with the robot and computer. These guides permit further user enhancement of the system. In fact, ToT's microcomputer, the SYM 1, is already in widespread educational use. Student enhancements bring it to the peak of its powers.

ToT's speech synthesizer, the Votrax Type 'n Talk, also comes with its own manual. This device generates and organizes speech phonemes, and its operation actually offers instruction in the structure of speech. Words are programmed as characters in ASCII code, either as letters or as phonetic sounds. The vocabulary is unlimited.

Perhaps the most interesting educational feature of ToT is the on-board computer's interface with ToT's ultrasonic ranger. The ranger is ToT's eye on the world. Students learn how the ranger is monitored by the computer for either motion detection or obstacle avoidance. In turn the computer activates program subroutines in response to ranger input. These subroutines may be either speech or movement. The possibilities are extensive and stimulating to the imagination.

ToT's On-Board Computer

ComRo ToT's on-board SYM-1 computer has 4K. The five motors, one light, and speech output are all controlled by the SYM-1. No external computer is necessary, although ToT's RS-232C port may be interfaced with any 6502-based microcomputer for uploading and downloading memory—i.e., programs—and for direct toggling of ToT's motors from the mother computer. Compatible computers include Apple II and *IIe*, Commodore, VIC 20, and the Commodore Pet.

The robot's 16-key hexadecimal keypad permits a debug function with all register contents displayed on an LED panel. The user can trace the computer's handling of a program via this function. Other command keys include save and load functions which store and retrieve programs from ToT's built-in tape recorder. The computer is capable of block memory transfers.

The ComRo ToT Software Package

The ToT Software Package that comes with each ToT unit is designed to make the robot user-friendly by putting several interesting ToT application programs at your fingertips. The accompanying manual also explains how to write your own programs. The Software Package is a cassette which is inserted in ToT's own on-board tape recorder, then loaded into the memory of ToT's microprocessor. The cassette contains all necessary subroutines, plus:

- ☐ Ranger mode obstacle avoidance. This program enables ToT to wander about avoiding obstacles and people all on his own.
- ☐ Guard mode. This program enables ToT to stand guard over an area and challenge anyone who crosses his sensor field with some threatening gestures and the announcement that the intruder is "under surveillance."
- ☐ Greet mode enables ToT to greet anyone who similarly crosses his field of vision with a handshake and an announcement: "Greetings Earthling, nice to see you."
- ☐ Survey mode enables ToT to survey his surroundings announcing the distance in feet and inches to various objects in his environment.
- ☐ Time mode enables ToT to run a real-time software clock and announce the time when someone enters or crosses his field of view.
- ☐ Snooze mode enables ToT to take a nap, then wake up at a preset time and talk or perform some action.

☐ Command Override mode enables ToT to perform in one mode, then at a preset time go into another mode, then at a preset time go into another mode, etc.

It is not difficult to modify the speech and/or actions in these programs to your own specifications. The manual gives an explanation.

ToT Advantages

As you can see, the ComRo ToT is an excellent personal robot with many valuable features and attributes. To help you decide whether ToT is for you, let's tally up some of the advantages of owning this worthy electromechanical "person."

Upgradeable. New ComRosoft programs are under continuous development and are designed for use on all ToT units. New ToT hardware is designed to retrofit all ToT units. In other words, ToTs are upgradeable. Your ToT unit will not become quickly obsolete or be replaced by a new model.

Fully Programmable. Just enter the program and start it running. ToT will deliver your message and interact with people all on his own. This makes ToT ideal as a promotional, or show, robot.

Home Security Program. Guard Mode deters intruders.

On-Board Computer. When you buy a ToT robot, you have all you need to enjoy a personal or promotional robot. No home computer or expensive software is necessary.

User-Friendly. No computer experience is necessary to operate ToT. Enough software comes with ToT to enjoy the robot without doing any programming. When programming is desired, step-by-step instructions are provided in the user's manual.

Extremely Versatile. With Dual Control Capability, two operational arms with plug-in accessories, four-wheel drive for crossing floor moldings, and human-scale strength for lifting and moving objects, ToT is one of the most versatile robots available. An optional gripper enhances its lifting capability.

Rugged, Durable Construction. Hand-built, hand-finished, ToT is tough. Molded, sturdy 3/32-inch resin-reinforced fiberglass, 3/8-inch acrylic plates and steel rod superstructure make ToT a most rugged robot.

Excellent Servicing System. *ToT-Line* is available for minor user repairs. Modular construction comes apart into head, body, and base units. These units may be returned to ComRo Inc. for servicing via United Parcel Service. Only the module with the problem need by returned.

Good-Looking. ToT has been photographed for international fashion magazines!

Buying ComRo ToT

Prices for this robot and its options, as of August, 1984, were as follows:

☐ ComRo ToT Standard, with On-Board Computer Control, and Accessories: $3,495.00
☐ ComRo ToT Dual Control, with On-Board computer control, plus handheld radio control: $4,495.00
☐ Optional Remote Control Gripper: $350.00
☐ Optional Remote Voice System: $350.00
☐ Optional Travel Case: $300.00

Accessories included with the Standard and Dual Control models include:

☐ Job Bar
☐ Payload Bucket
☐ Lifting Hook
☐ Carpet Sweeper
☐ Tray and Cups
☐ User's Manual
☐ Free Subscription to *ToT Times*

Inquire about special customizing for your unit or custom robot projects. ComRo ToT is available exclusively from the manufacturer.

Orders are initiated by sending in a 20 percent deposit together with customizing instructions: color, whether for home or promotional use, etc., including customized programming. ToTs come with a 90-day service warranty and are fully tested prior to shipment. The units disassemble into three sections for easy return of individual modules for servicing. A 24-hour ToT-Line number is available for service and operating advice.

HUBOTICS

HUBOT is unique among the panoply of personal robots. His manufacturer, Hubotics, says that HUBOT isn't designed specifically for educational purposes, as a toy, or solely for entertainment. Instead, HUBOT is acclaimed as a home appliance: the "ultimate home appliance of the 80s," according to Michael Forino, president of Hubotics.

Forino says that he and his management team actually went to its market—upscale and affluent consumers—and asked them what they wanted a home robot to do. What they found was that the consumer wanted more, much more, than a mobile computer workstation and fun and games. They wanted a robot that would be "Mr. Everything;" in effect, a multipurpose appliance.

Hubotics believes that HUBOT fits the bill. The robot is designed as a combination servant, teacher, entertainment center, and mobile computer. He is also made to be expanded and enhanced so that he will not become obsolete as technology continues its inexorable march.

HUBOT's Grand Entrance

May 1, 1984 marked the arrival of HUBOT on the world scene. The mass-produced appliance, first shown at the January, 1984 Winter Consumer Electronics Show, was formally introduced to consumers at Abraham & Strauss stores throughout New York, New Jersey, and Pennsylvania. His premier showing was accompanied by an active following of fascinated shoppers—some awe-stricken, others amused, and still others shocked. After all, most had never seen a walking, talking, live robot with a television screen face.

Store personnel and customers were amazed as HUBOT moved through A&S aisles on his own battery-charged motor and stopped to talk along the way. The 41-inch tall, 125-pound HUBOT gave continuous demonstrations using his powerful HuBrain CP/M® -compatible computer and his SpeakEasy

HUBOT, shown here without his optional arm, is touted by his manufacturer as a do-it-all robot designed specifically as a home appliance (courtesy Hubotics, Inc.).

language to show customers how he could be smart, fun, and mobile. HUBOT also entertained customers with his full-channel, 12-inch, black-and-white television, his Atari video games, and his AM/FM stereo cassette tape deck.

Since his grand entrance into the marketplace, HUBOT has gone on to achieve considerable marketing success. Broadway and Wienstock's department stores in California began offering him to customers, as did department stores and specialty shops in such geographically varied locales as Harrisburg, Pennsylvania, Minneapolis, Minnesota, and Ft. Lauderdale, Florida. In addition, the popular upscale *Sharper Image* catalog now offers the home robot to its mail-order customers.

The Hubotics Company

The HUBOT home appliance robot is the first product offered by Hubotics, Inc., but the company says it will be only the first of many automated home products and appliances. The company believes this market has barely been tapped and that it is time to apply what has been learned in the factory and office to the home.

Before too long Hubotics expects home automation to be built right into houses as they are constructed, including communications lines and computer functionality. The HuBrain computer now in HUBOT the robot will be a nucleus for many planned products. For example, the HuBrain computer will be adapted to fit right into the walls of new homes as a central controlling device. A major application will be energy management, not only for heating but also for appliances and lights. Other uses include sentry duties, automatic timer for many devices, home computer, and more.

Other existing technology will be creatively applied to other areas of the home for added convenience and increased productivity and enjoyment. Robert Sachs, vice president for sales, says that the lawn mower or even the golf cart could be the next items to be automated. If the initial success of HUBOT the robot is carried over to other consumer household appliances and products, the Hubotics Company may be off to the races.

The driving force behind the Hubotics Company is its co-founder and president, Michael N. Forino. Previously director of customer services at International Robomation/Intelligence, Forino is reputed to be a wise entrepreneur with good instincts for the high-tech market. One of his previous enterprises was a firm called Spacetronics, a security alarm company.

Hubotics Company recently moved into its new facilities, a 10,000-square-foot building in Carlsbad, California, which houses all of the company's administrative, manufacturing, and marketing operations.

A Look at HUBOT

HUBOT costs more than many other robots; his basic price tag is $3,495, and that doesn't include a wide assortment of options. Robert Sachs contends, however, that HUBOT is well worth the asking price because the owner gets so much capability and function. A look at HUBOT's impressive attributes does lead you to believe that this is a vastly superior machine-person.

Expandability. Unlike other household appliances, HUBOT can be expanded or enhanced, which overcomes any problems of obsolescence and also allows a purchaser to buy a basic model and add options as desired. He is like a household erector set. Even the computer boards can be swapped out when more powerful microprocessors are desired.

Intelligent. Built right into his body is a special computer called the HuBrain. Like the human brain it has two parts. One side controls HUBOT's robotics. The other controls all of HUBOT's functions and is a full personal computer with the CP/M operating system, giving him the ability to run a wide variety of popular applications. The SpeakEasy language makes it simple to command HUBOT. With only the push of a button or two, HUBOT will do whatever his owner requests.

HUBOT is a full personal computer mobile workstation. He has a detachable keyboard that pulls out for ease of use, a built-in, 80-column monitor, and even a space for the user's toes when he sits at the robot. Other features include a 5 1/4-inch floppy disk drive, an optional built-in printer, and a standard communications link so HUBOT can tie into other computers and information services.

Fun. HUBOT is packed full of entertainment. With unlimited vocabulary, HUBOT has the ability to say almost anything in a synthesized voice. When

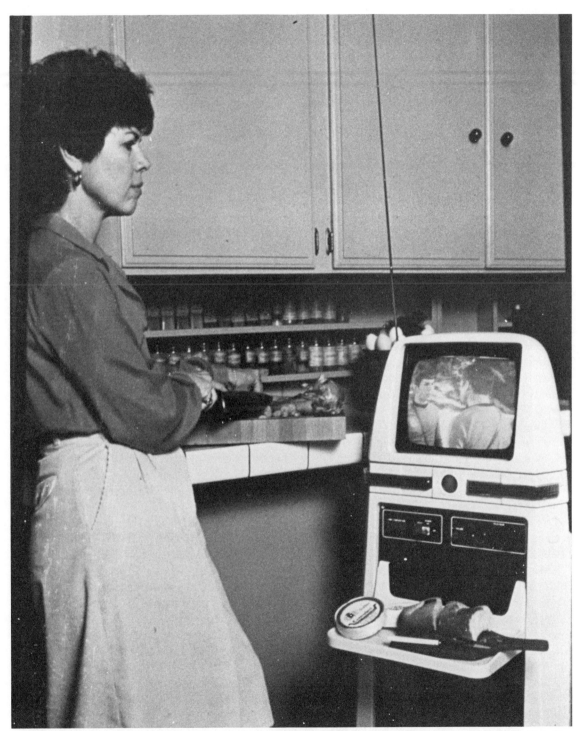

HUBOT is a functional household robot that appeals to housewives, working parents, and children alike (courtesy Hubotics, Inc.).

HUBOT talks, his face on the monitor moves.

An optional drink tray is available. Flashing red collar lights give HUBOT a bit of Hollywood pizzazz.

Mobility. HUBOT can be taught to move along designated paths. Once taught, HUBOT can move along the same path again with only the push of a button. If something gets in HUBOT's way, he will stop and verbally inform his owner. A rotating Obstacle Sensing Processor (OSP) collar detects objects in his way. HUBOT will soon be able to navigate around obstacles. HUBOT can also be made to move by using a typical video-game joystick.

Owners can tell HUBOT to turn on the radio, television, or other functions through the computer, rather than doing it themselves. In the near future, a voice command option will allow HUBOT owners to verbally instruct HUBOT to perform these functions or to move.

Construction. HUBOT was designed to aesthetically please in the home as well as to be rugged and durable. The 44-inch-tall HUBOT is primarily beige and has only rounded edges. The HuBody comes from a single mold of polyethylene plastic. The unibody construction combined with this tough material makes HUBOT nearly indestructible. His body can't be scratched or even broken by a hammer blow. Even the monitor has a special second screen, making it almost impossible to break.

HUBOT has wheels similar to a lawn mower so he can move safely over almost any terrain. His balance is so good it would be very difficult to accidentally knock him over.

Miscellaneous. Miscellaneous features include a digital clock displaying time and temperature. HUBOT's battery can be recharged by merely plugging him into an ordinary wall socket.

Options. HUBOT already has several optional packages, and many more will be available soon. Optional enhancements are easily adaptable to the existing HUBOT package.

Current options include: a protective cover, made of tough cordura fabric, which keeps HUBOT dust-free when he is not in use; a homebase, or plastic plate, which gives owners a reference point for use in teaching HUBOT paths and directions in the home; a drink tray, which can be used for carrying various items; and an on-board, 40-column, matrix printer.

Future options will include: a voice command module, to enable HUBOT to respond to verbal commands; obstacle recognition and navigation, which will enable HUBOT to recognize an obstacle in his path, navigate around it, and then continue on his predetermined route; and sentry package, which will enable HUBOT to act as a fire and burglar alarm. Dexterity—HUBOT will be fitted with an arm and hand so he can pick up and carry objects. Autocharger—HUBOT will detect when his power is low, automatically go to a wall unit, and recharge his battery. Light-duty vacuuming; remote control to give the user the ability to remotely issue commands to HUBOT from anywhere in the house; and a remote telephone and auto-dial; also available will be a smart servant module, a small electronic box that can be plugged into the wall socket, which will act as the control station for home appliances. After a verbal command, such as "Turn on the dishwasher," is given to HUBOT, he will send a signal to the Smart Servant Module, which will automatically activate the dishwasher. Enhanced control will also be available. A grid diagram of a house can be shown on HUBOT's monitor and his path to anywhere in the house determined by moving an image of HUBOT on the screen through the grid.

Buying HUBOT

To get your own robot butler and household helper, you'll have to shell out $3,495. If you order direct from the company, shipping costs are extra (about $20). That's a lot of money, but you'll get a first-rate robot that comes with a full 64-key keyboard, a Kaypro-like CP/M, 8-bit computer with 128K, 80-column screen, a built-in black-and-white television set, an Atari 2600 video game player, and a host of other goodies, Plus there's voice synthesis, so that when you get HUBOT home, you peel off the sticker on top of his head, turn on the key, type in your name, and hear the robot exclaim, "Hello John, my name is HUBOT. How may I serve you?"

If all this isn't enough, you may be enticed into buying one or more of the options soon to become available to make HUBOT even more useful, educational, and fun to have around. Here's a quick read-out of tentative prices of major future options:

☐ Second disk drive: $395.00

☐ Articulating (flexible) arm to pickup objects: $700.00

☐ Vacuum cleaner attachment: $300.00

☐ Sentry fire and police alarm system: less than $300.00

☐ Smart Servant to control household appliances and lights: less than $150.00

Table 2-4. Product Specifications for HUBOT.

Model: HUBOT PREMIER 1000

FEATURES	**SPECIFICATIONS**
HuBrain Computer	Z80A, 64K user RAM
CRT Display:	
Text	80 columns/24 lines
Software	CP/M 2.2
	SpeakEasy (Instruction Language)
Keyboard	64 key ASCII/detachable
Disk Drive	5 1/4″ floppy DS/DD
Interface Panel Ports	2 - joysticks, 1 - RS232,
	1 - parallel/centronics
	1 - battery charger
Dimensions	22″ in diameter × 45″ height
Weight	110 lbs. approximately
Motor Drive Speed	Approximately 12″ per second
Construction	HuBody; UniBody polyethylene
Power Supply	12V 40 amp-hr. sealed lead acid battery
Battery Charger	12V, auto/shut-off; operates on
	110VAC 50/60 Hz.
Entertainment:	
Television	12″ black and white
Radio	AM/FM stereo, cassette, equalizer,
	dual speakers
Games	Atari cartridge capability, 2 joysticks
Voice Synthesizer	unlimited vocabulary (text to speech)
Safety:	
OSP Collar	Sonic transducer, collar indicator
(obstacle recognition)	3 infrared transducers

OPTIONAL FEATURES	**SPECIFICATIONS**
Convenience Package	Serving tray, Hubot cover,
	HomeBase
Printer	40 column, dot matrix
2nd Disk Drive	5 1/4″ floppy DS/DD
Coming Soon:	
Voice command	Real voice recognition for issuing commands
Dexterity	Articulating arm
Sentry Package	Heat/smoke/intrustion alarm, 360° obstacle recognition
Auto Recharger	Automatic battery charging station for Hubot
Vacuuming	Light duty, dual roller
Remote Command	Issue commands remotely
Telephone	Remote telephone/300 baud modem
Smart Servant	Household appliance control system

An option already available is the on-board, 40-column, dot-matrix printer that prints out data from the video screen. Cost is $300.00

HUBOT comes with a free, basic premier protection plan that provides the owner with 1 year free replacement of parts, 30 days on-site labor, and 90 days free software updates and technical support through a toll-free number, 7 days a week. Additionally, HUBOT owners have the choice of purchasing extended protection through optional plans. Option I extends on-site labor for 11 months; Option II provides the owner with continuing software updates, a 300-baud modem for communication with other computers, electronic bulletin boards, mail systems, technical support, and on-line diagnostics.

HUBOT is available primarily through major department stores and retail shops (other than computer stores). He can be ordered directly from the company if a participating store is not accessible in your area. To obtain a list of stores and shops that offer the robot or to order direct from the company, write or phone Hubotics.

TOMY

If you're not quite ready for an expensive, do-everything robot like HUBOT, with its keyboard and on-board sophisticated computer, or HERO I, with his impressive educational talents, then Tomy's three fascinating robots—DINGBOT, VERBOT, and OMNIBOT—just may be right for you. Tomy Corporation, one of the world's largest manufacturers of toys, has brought tomorrow's technology to today's world of play. Robot technical afficionados and hard-core advocates may insinuate that Tomy's machines are "mere toys." The fact is, however, that this trio promises to deliver amazing versatility and superb joy to robot lovers interested in a low-cost, fun robot for pleasure and entertainment.

True, there are liabilities in Tomy's robots, and their performance may not be ideal for the serious roboticist. For one thing, even OMNIBOT—Tomy's top-of-the-line robot—lacks the ability for computer interfacing. His limited programming is built in and is not expandable. OMNIBOT does, however, pack a remote-control capability and an LED alarm clock; he transports items on a tray, moves in all four direc-

tions, and speaks through remote microphone or from a built-in cassette tape player, all for under $300! These robots aren't R2D2 or the androids you saw in the acclaimed science-fiction thriller *Bladerunner,* but they are a great way to begin your journey into the world of robotics.

One thing is for sure: Tomy will be selling a huge quantity of these robot creations in coming months and, perhaps, years. The marketing people at Tomy have proven remarkably adept at picking up on public moods, fancies, and trends. They evidently see personal robotics as a dynamic, up-and-coming field. Thus, they are busy putting their considerable marketing savvy—and a lot of dollars—into selling their robots.

From DINGBOT to OMNIBOT

Standing a mere 5 inches high, the adorable DINGBOT is perhaps the most lovable robot in the Tomy lineup. With a map that only he can follow, DINGBOT skitters along, bumping into objects along the way. At each encounter, DINGBOT stops, chatters to himself as his head turns from side to side appearing to be reading his map, checks his bearings, and moves along. Battery operated—and a bit of a comedian at heart—DINGBOT is recommended for youngsters 4 and up.

Next in the Tomy lineup is VERBOT, a voice-activated robot that responds only to your voice and secret commands. By entering your commands into VERBOT's memory through a wireless remote-control microphone, VERBOT can be programmed to perform eight separate functions. Upon verbal command, he will move forward, backward, turn left or right, pick up a small object and carry it or set it down, and stop. When ordered to "smile," VERBOT's eyes light up and he emits a cheerful clicking sound. VERBOT is battery operated, with a remote-control microphone transmitter. He is recommended for youngsters 6 and up.

Standing nearly 2 feet tall, and high-tech in design, the third entry into Tomy's robot line is OMNIBOT, a preprogrammable, functioning robot with far greater capabilities than his two siblings.

OMNIBOT comes complete with an on-board microcomputer and cassette tape deck, a remote-

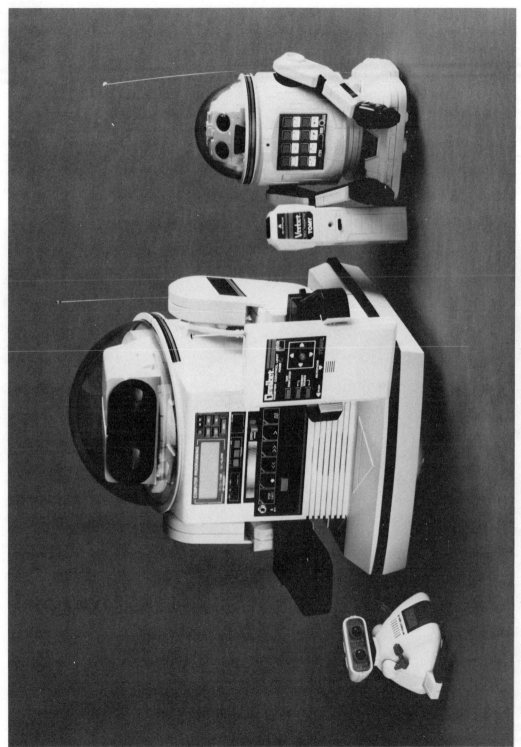

Tomy introduces "tomorrow's technology at play" with (left to right) DINGBOT, OMNIBOT, and VERBOT (courtesy Tomy Corporation).

control transmitter, digital alarm clock, built-in microphone, and a manual grasping hand. Using the remote-control transmitter, you can program OMNIBOT to repeat up to seven different programs at designated times up to seven days in advance. OMNIBOT can be programmed to perform by remote control or from memory. He moves forward, backward, turns left or right, and will stop on command. By remote control, OMNIBOT will repeat a series of commands when programmed, or repeat those same maneuvers at a later time through the on-board digital clock/tape memory system.

OMNIBOT can deliver your written message in his manual grasping hand, or your pretaped message via his on-board microcomputer and tape deck. He can also serve as your personal butler, delivering canapes at a party from his detachable serving tray; bring you the evening paper and your slippers after a long day at the office, then retreat to a corner of the room and play your favorite cassette tape; or walk into your room in the morning at a predetermined time and wake you with a prerecorded message. Battery operated, with a built-in rechargeable battery and recharging unit, OM-

OMNIBOT, the preprogrammable electronic robot from Tomy, may not be as sophisticated as some others, but he's remarkable in many respects--and the price is right! (courtesy Tomy Corporation).

Ver-bot™

- Voice activated.
- Responds to commands entered into memory through wireless remote control microphone.
- Performs 8 tasks.
  - Move forward.
  - Stop.
  - Move backward.
  - Left turn.
  - Right turn.
  - Pick up object, carry it.
  - Put down object.
  - Smile (eyes light up).
- Batteries required.

Omni-bot™

- Operates under radio control.
- Repeats a series of commands when programmed.
- Can repeat these functions at a later time through on-board clock/tape memory.
- Can play any standard cassette.
- Capabilities.
  - Speak and deliver messages through on-board tape unit or through remote microphone.
  - Built in alarm clock.
  - Carries objects with detachable tray.
  - Manual grasping hand.
  - Move forward.
  - Move backward.
  - Stop.
  - Right turn.
  - Left turn.
- Built in rechargeable battery and recharging unit.
- Batteries required for tape player and clock.

Ding-bot™

- The most loveable robot.
- Changes direction whenever object is encountered.
- Sequence as follows:
  - Walks forward, encounters object.
  - Stops. Head turns from side to side.
  - Chatters with a chirping effect.
  - Turns right, moves in new direction.
- Comes with his own map only he can follow.
- Batteries required.

Tomy's Tantalizing Trio (courtesy Tomy Corporation).

NIBOT is recommended for ages 9 and up.

Buying the Robots

Each of Tomy's three robots come complete. The purchase price includes all accessories and an instruction manual. OMNIBOT retails for about $295, VERBOT around $55, and the diminutive DINGBOT for just $10.

Tomy has such a reputable name that its robots should be in thousands of retail outlets around the nation, from Montgomery Ward (including Wards catalog) to Toys R Us. If, however, you do experience difficulty in locating a retailer, you may write to Tomy Corporation.

IDEAL (CBS TOYS)

MAXX STEELE™ is his name. He walks, he talks, he plays games. He has a fully articulated,

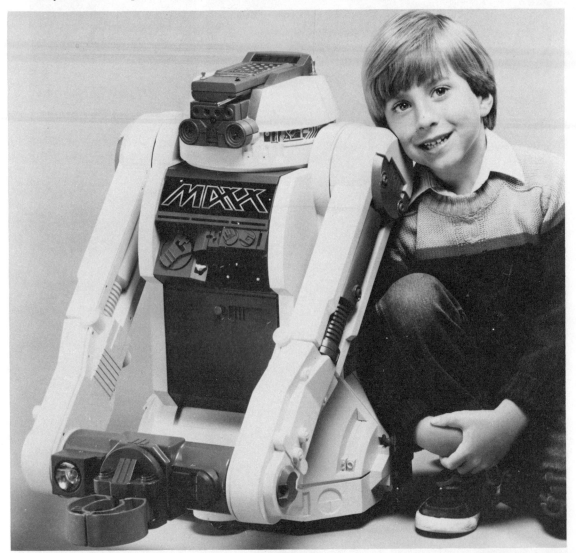

Program MAXX STEELE to speak and move. This 2-foot-tall electronic robot from Ideal can be controlled either by radio or by his programmable memory system. MAXX is mobile, can pick up objects, and can even play electronic games with the whole family. (courtesy Ideal—CBS Toys).

It's the good guys vs. the bad guys with ROBOFORCE, a band of robot action figures from Ideal. MAXX STEELE leads six good robots against HUN-DRED and his four bad guys. Special design, features, and weapons give each of the ten his own personality (courtesy Ideal–CBS Toys).

motorized grasping claw; moving arms and wrist; excellent mobility, storage, and carry compartments, and an electronic computer brain. MAXX is the leader of ROBOFORCE™, a warrior robot team that is unparalleled in imagination and derring-do action.

MAXX STEELE is a product of one of America's largest toy manufacturers, Ideal, a subsidiary of CBS Toys. MAXX has hit the ground running, being mass-produced and offered in department and toy stores and specialty shops all around the nation, as well as in foreign countries. As a personal robot, MAXX comes in three versions: as a fully assembled programmable robot, as a remote-control erector kit, or as an inexpensive ($6), but necessarily limited, toy figurine. Each of these robots goes by the name MAXX STEELE, but each is significantly different in capability, and to some extent in physical appearance.

The ROBOFORCE

To complement MAXX, the folks at Ideal offer to the world 10 action figures: tiny electromechanical and motorized robotic figurines that are much like Tomy's DINGBOT robot. These 10 action figures are made up of 6 "good guy robots," led by hero MAXX STEELE, and 4 bad guys, led by HUN-DRED. All are 5 3/4-inch tall toys with real working movements.

These 10 ROBOFORCE robotic figures are certainly not true personal robots as we generally define that term, but as toys they are exciting and unusual. Just consider their well-conceived names: there are bad guys: WRECKER, the demolisher; ENEMY, the dictator; CRUEL, the detonator; HUN-DRED, the conqueror; VULGAR, the destroyer. On the good-guy team we find: SENTINEL, the protector; S.O.T.A. the creator, and of course, hero MAXX.

The 10 ROBOFORCE action figures have air bellows for the arms, which flex into action when a button is pushed on the robot's back. A suction-cup base enables each robot to stick to surfaces, upright or upside down, or lift flat items. Specialized design, features, and weapons give each of the 10 a separate personality. Kids aged 4 and up will

want to join forces with them all.

Add to the adventure with three separate ROBOFORCE air-powered vehicles: ROBO-CRUISER (for the good robots), a robot vehicle that blows up balloons with a special air nozzle; DRED CRAWLER (for the bad) and the COMMAND PATROLLER vehicle and playset (for both forces). THE FORTRESS OF STEELE, a trilevel playset environment, offers hours of robot action with features including a crane, revolving weapon mount, fold-out sections, sliding doorway, control panels, throne room, prison cell, and accessory storage areas. The entire set is a case that folds up for portability. A comic-book style storyline/cross-sell booklet comes with each robot, vehicle, and playset. All require AA batteries.

Meet MAXX STEELE

MAXX STEELE is a 34-pound, 2-foot tall, electronic programmable robot which can be controlled either by an accompanying radio or his programmable memory system. MAXX is talented; he has a moving arm with a fully articulated claw that's nearly as flexible as a human wrist. He is mobile, can pick up light objects, speaks 20 different preprogrammed phrases in addition to a 150-word vocabulary, and can play electronic games with his owner. In addition, he sports LED displays to read out his program or the time of day from his internal clock/timer, music overlays for his programmable keypad, a searchlight in his wrist, a built-in shelf compartment to carry objects to his owner, and a rechargeable battery pack and transformer. What's more, MAXX is modular and expandable to accept future accessories and new hardware and software developments.

Build Your Own Working MAXX STEELE

For those wishing to build their very own companion, there is the ERECTOR MAXX STEELE remote-control robot. Owners can send MAXX forward, back, left, and right. His three-wheeling base provides total maneuverability. His arms raise and lower, and one arm has a claw hand that opens, closes, and rotates to pick up objects. The claw arm

Set robot hero MAXX STEELE in his air-powered ROBO CRUISER and he's ready to lead the ROBOFORCE into action. Also available from Ideal is the DRED-CRAWLER, a vehicle specially designed for HUN-DRED and his band of bad-guy robots (courtesy Ideal—CBS Toys).

is activated by a pistol grip and lever located at the shoulder end and is compatible with any other ERECTOR model. the other arm has a magnetic disk to pick up steel parts. This model of MAXX STEELE isn't as versatile—he can't talk, for example—but costs much less.

Buying MAXX STEELE

Here's the approximate retail price for MAXX STEELE, the ROBOFORCE team, and their vehicles:

- ☐ MAXX STEELE (programmable robot): $350
- ☐ MAXX STEEL (erector kit): $50
- ☐ ROBOFORCE action figures: $6 each
- ☐ ROBOCRUISER and other vehicles: $12 each

Incidentally, if you want to keep up with robotics and other high-tech subjects, check out *ROBOFORCE* magazine, a slick publication available at many bookstores and supermarkets. Designed for children, and nontechnical in presentation, the articles are interesting enough that most adults will enjoy the magazine.

Ideal and CBS Toys have nationwide distribution; so all major toy stores and many department stores will carry the ROBOFORCE line.

ANDROBOT

Ask 10 knowledgeable people to name the most well-known personal robot, and chances are at least 9 will answer, "TOPO." Ask the same 10 people to name the man responsible for TOPO, and without hesitation most will reply, "Nolan Bushnell." Nolan Bushnell and his first creation, TOPO, certainly de-

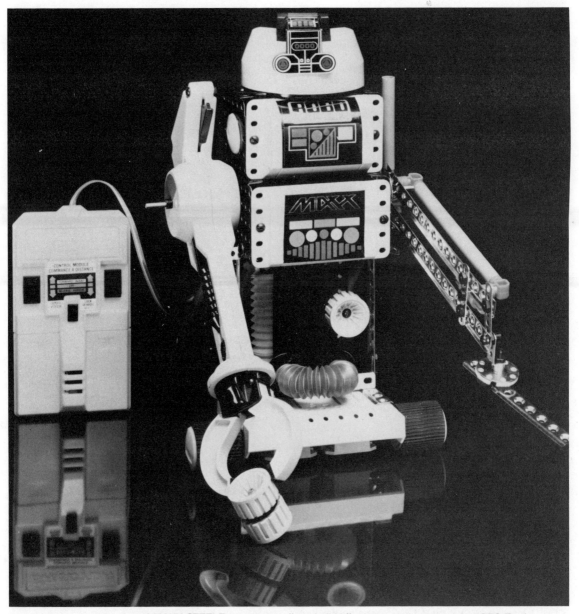

This remote-control version of MAXX STEELE comes as a "you build-it" erector kit (courtesy Ideal-CBS Toys).

serve this widespread acclaim, for it is this pair that was the chief catalyst for the now rapidly expanding personal robot market. This is notwithstanding the competing claims of others who hold up Heathkit with its HERO I, RB Robot Corporation with its RB5X, or any of a half dozen other entrepreneurial companies as the one company that got the personal robot industry off and running.

Nolan Bushnell and Androbot will undoubtedly be fondly remembered and be given credit in the future annals of robotic technology. Bushnell gained fame and fortune well before he began Androrobot, Inc., the company which is responsible for TOPO and which now offers an entire family of personal robots. In 1972, the bold, brilliant entrepreneur developed one of the first computer video games and went on to found Atari and lead that firm into its glory days. Later, after he had sold his interest in Atari, the company began to falter. Nolan also founded Chuck E. Cheese's Pizza Time Theatre, another company that zoomed to prosperity in record time.

Lately, the intrepid Bushnell has seen his winning image tarnished as first the Pizza Time Theatre chain, then Androbot fell on hard times. Androbot, for example, has gone from nearly 300 employees in 1983—including a number of first-class engineers and computer scientists—to less than 20 employees today. This reduction is, however, more a reflection of Bushnell's zealous belief in the future of the personal robot industry than an indicator of faulty products or management savvy. Bushnell and Androbot simply overestimated the depth of the market.

Initially, sales of Androbot's robots were far below the company's somewhat grandiose expectations; however, the truth is that Nolan Bushnell was probably just a few years ahead of his time. The market is only now beginning to show signs that it may indeed match up to his bold vision of boom and peak demand. In any case, Androbot has retained a nucleus of personnel at its headquarters in San Jose, California, the heart of Silicon Valley, and is still very much in business. Some of the early bright optimism has dimmed, but it is possible that the future will see the company and its products vindicated.

It is important to consider just what Androbot has to offer in the way of personal robots. Looking over its four-robot family, it is accurate to say that these are exceptionally fine products. Described by one computer and robotics authority as resembling geometric snowmen in shape and appearance, each of the four is unique and designed for a special segment of the market: entertainment, toy, educational, and home companion. In addition, the robots are priced to fit a spectrum of pocketbooks.

The name for Bushnell's company, Androbot, is a simple but clever combination of *android,* meaning humanlike, and *robot.* The four robots offered by the company are, in fact, vaguely humanlike in appearance, and they are attractive. They also embody a wide-ranging array of technological sophistication and wizardry. Let's cover each robot in turn, starting with the least expensive, AndroMan, and working up to B.O.B., the most impressive of the quadruplet.

AndroMan

AndroMan is a sophisticated, 12-inch-tall, mini-robot and a real-life video game set designed specifically for the Atari VCS 2600. A video game cartridge supplies action on your TV screen, and an adapter module lets you control AndroMan with a joystick, using an advanced two-way infrared data link. AndroMan is exciting because he is the first real-life robot and video game combination.

Each game includes realistic dimensional playing pieces you arrange on a full-size Gamescape. Play alternates between your video screen and your real-life AndroMan.

Your objectives are to manipulate AndroMan, step by step, through the 3-D Gamescape, before you can proceed through ever challenging video play. Then, while you're deep in concentration on the screen, AndroMan talks to you—warning, encouraging, or cajoling.

F.R.E.D.

As a junior member of the Androbot family, F.R.E.D. is an educational introduction to the world of personal robots. His size is small (12 inches), and his stature is squat, but his abilities are large.

F.R.E.D. (Friendly Robotic Educational Device) can be on the go almost anywhere you choose—on the floor or on a tabletop. He can be controlled by his portable remote infrared controller or via your personal computer keyboard or joystick. When you use your computer and his software to guide him, he constantly updates your computer on his progress. An optional "teach mode" lets him remember his travels and duplicate those moves on command.

A drawing pen attachment lets you turn F.R.E.D.'s artistic talent loose. While you create patterns on your computer screen, he'll translate those designs into precise drawings.

F.R.E.D. (Friendly Robotic Educational Device) has a drawing pen attachment. Using your computer keyboard and screen, or a controller, you can guide the robot in drawing various shapes (courtesy Androbot, Inc.).

A voice synthesizer with a 45-word vocabulary gives F.R.E.D. literally hundreds of communications phrases. He's designed so that new software can be added as it becomes available. F.R.E.D. has sensors that aid him to avoid falling off the edge of a table or tripping down the stairs.

TOPO III

TOPO I was the earliest personal robot introduced by Androbot, and his capabilities were minimal. The upgraded TOPO III model is another matter, however. Deanna Ladd, Androbot's sales manager, explains that TOPO III is designed to be the ultimate computer peripheral: a friendly extension of your home computer. "But TOPO III is much more than just a home companion," Ladd says. "He is ideal for educators and hobbyists to study geometry, the metric system, robotics, and computer languages such as Forth, BASIC, and Logo."

A sophisticated machine, TOPO can talk or sing, walk with *Andromation*—his two-wheel-drive system—and spin and dance. He also comes with many options, including software packages, joystick, and heat and sound sensors. TOPO also has programmable speed and acceleration rates, and there's a panic switch on his head you can activate if you want the robot to cease whatever activity in which

Table 2-5. Specifications for TOPO III.

Supplied Hardware

TOPO III ROBOT

* (2) 8031 Microprocessors
  - #1 Controls speech and communications
  - #2 Controls motion
* Two-way Infrared Communications
  -Communication range of 30 feet (line of sight)
  -Receives commands from Base Communicator and reports back status
  -Automatic error detection and message acknowledgment
* Speech Synthesis System
  -Text-to-Speech mode for simple operation
  -Phoneme mode for more complex applications
      -Foreign language
      -Singing
  -Adjustable pitch, rate, and volume (both modes)
* Motion
  -Moves forward, back, left, right, and in smooth arcs
  -Programmable speed and acceleration
  -Simple motion control without computer, using built-in headswitch
* Expandable Card Cage (for future accessories and features)
* (2) 12-Volt Industrial Grade Batteries
  -Operates TOPO for up to three hours
  -Rechargeable with 24 volt charger (supplied)
  -Battery monitor circuit provides over-charge protection
  -Alerts operator of low battery condition
* Heavy Duty Motors and Gear Boxes
* High Impact, 22 piece plastic body of fire retardant ABS
  -Footlights indicate direction of motion
  -Armlettes allow TOPO to carry very light loads
* FCC/UL Approved

BASE COMMUNICATOR
* 8031 Microprocessor
  -Up to 9600 Baud communication with the computer
  -1200 Baud communication with TOPO
* 12 volt power adapter
* RS-232 cable

SUPPLIED SOFTWARE

OWNERS MANUAL
* Easy to read
* Easy to understand
* Lots of examples
* Step by step from the simple to the complex

DISKETTE
* Your choice of either
  -TOPOSOFT for the Apple (FORTH)
  -TOPOBASIC for the Commodore (BASIC)

REQUIRED EQUIPMENT

FOR APPLE COMPUTERS
* Apple II, +, or IIe
* 48K RAM
* Apple Super Serial Card (RS-232)
* Diskette drive

FOR COMMODORE COMPUTERS
* Commodore 64
* VIC-1011 (RS-232)

* Diskette drive

he's engaged. TOPO is armless for now, but arms may become available.

You Command, TOPO Performs. Once you've acquainted TOPO with his new home, a simple computer command or joystick movement will start him off and running. For instance, while you're in the kitchen, the keyboard command "TOPO to patio" will send him scurrying over a previously memorized route to serve drinks to guests from his optional Androwagon. A wireless infrared communications link relays information between TOPO and your computer throughout your house. TOPO software is now available for Apple, Commodore, and VIC 20 computers.

Multitalented TOPO You can direct TOPO to speak in many languages with his optional programmable voice. Make him sing by controlling the pitch, volume, and speed of his speech.

With TOPO's optional joystick "teach mode," you'll maneuver him around the room while the computer saves the list of commands being executed. Later, you can replay the action and voice routines and watch TOPO do his stuff again and again. He doesn't forget.

B.O.B.

B.O.B. (actually B.O.B./XA, which stands for Brains on Board Expandable Androbot) is Androbot's most advanced and most expensive personal robot. Unlike F.R.E.D., AndroMan, and TOPO III, all of which require support from an external computer, B.O.B. is an autonomous robot, and a powerful one. He is outfitted with enough sensory capability and processing power to navigate in a home or office environment. In addition, the optional AndroLift is a device which allows B.O.B. to reliably pick up and move loads up to 18 pounds using only his self-contained sensors.

The 50-pound robot, which is practically a look-alike of his weaker brother, TOPO III, is easily programmed using a home computer for a terminal and file storage. B.O.B. comes with Forth, a language with integrated interpreter and editor. All robot operations (GO-CURVE, GET-SONAR-DISTANCE, SAY) are English-like Forth words. The sensors include ultrasonic range finders, curb feelers, a bar code reader, and an electronic compass.

B.O.B. has been designed with expandability in mind. The metal frame has several locations for additional sensors or other accessories. The CPU board has eight IBM PC-compatible bus connectors for option boards and eight sockets for ROM software cartridges. The operating system has been designed to automatically configure itself for all software cartridges. Androbot is planning many future accessories for the robot.

A close look at B.O.B.'s powerful anatomy reveals his many credentials. Note that his "brain" is the powerful Intel 8088 chip.

Processor: 8088 CPU.

Memory: 64K to 256K RAM on CPU board, up to 128K ROM on CPU board (eight cartridges). More RAM and ROM can be added in bus slots.

Bus: Eight IBM PC-compatible slots.

Drive System: Two drive wheels with feedback control system for precise positional control and two casters.

Sensors: Ultrasonic rangefinders, infrared NavCode readers, electronic compass.

Speech: Text-to-speech and prerecorded vocabulary using Texas Instruments 5220.

Software: Operating system includes control of motion, sensors, and speech. Forth language with robot extensions, editor, and 8086 assembler also included. BobTerm program for home computer communication and file storage. (Requires IBM PC or Apple IIe computer for file storage and editing.)

B.O.B. has been criticized for his lack of a true arm. His is a simple, nongripper-type lifting device that operates only in the Y axis (up and down). Still, if having an IBM PC-compatible robot that walks, talks, and has an impressive sensory package sounds good to you, you'll find this robot a worthy possession.

Buying Androbots

The approximate retail prices for the F.R.E.D., TOPO III, and B.O.B. robots, plus prices of some of their accessory products, follow. Androbot has not yet made AndroMan available to the market place, and no price has been set. A warranty of 90 days

B.O.B. (Brains On Board) comes with a noteworthy package of sensory devices and has an on-board computer using the powerful Intel 8088 CPU (courtesy Androrobot, Inc.).

labor and 1 year on parts comes free with each robot purchased.

B.O.B. Standard configuration: $3,995. Includes robot with CPU, drive system, key pad, speech, four ultrasonic rangefinders, infrared Nav-Code reader, electronic compass, IBM PC or Apple *IIe* interface diskette, and cable.

B.O.B. Basic Model: $2,495. Without full features. Includes robot with CPU, drive system and keypad.

B.O.B. AndroLift System (Accessory): $495

TOPO III robot: $1,595. Includes Infrared base communicator, cable and chargers, Commodore 64 or Apple *IIe* software diskette and manual.

TOPO III Software Packages. One is included at no charge with each TOPO III ordered. Order separately only if support for more than one type of computer is desired.

☐ TOPOSOFT (Apple): Software package which permits the operation and programming of TOPO III on the Apple II line of computers in Forth. Requires Apple Super Serial Card and one diskette drive.

☐ TOPOBASIC (Commodore 64): Software package which permits the operation and programming of TOPO III on the Commodore 64 computer in BASIC. Requires VIC-1011 interface and diskette drive.

F.R.E.D. Robot: $350. Includes speech, infrared communications, pen attachment, and transmitter. Software packages are optional.

Some department and computer retail stores offer Androbot products. You may also order direct from the company, which will ship the order to you via truck or air freight (software by United Parcel Service).

PERSONAL ROBOTICS

I first learned of the exploits of RoPet-HR when, on May 22, 1984, I heard on the network TV evening news that the robot had testified that day before the United States House of Representatives as an authority on high technology. I then got in touch with Chris Skottegard, president of San Jose-based Personal Robotics Corporation (PRC) to find out more about his company and its robots—in particular, the up-and-coming RoPet-HR.

What I have found out is that PRC is an excellent young company which produces technologically superior products. RoPet-HR is a case in point. This interesting fellow won "First Place—Commercial Division" in the competitive Second Tournament of Robots in Santa Ana, California, in April, 1984. Furthermore, the robot had won the "Best Overall—Commercially Built" category contest the year before at the same tournament.

PRC says that RoPet-HR and his bodyless brother, RoPet-XR, are designed to be easy to use, reliable, expandable, and useful. "RoPet-XR is especially designed for home security," says Skottegard. "He'll ask for the secret password after detecting an intruder (which he does by using and infrared detector). If he hears the correct password, he does nothing. Otherwise, he'll sound a loud alarm. With a modification to come soon, he'll even call the police. Automatically."

RoPet (both the HR and XR models) is entirely self-contained with an on-board Z-80 computer, speech generation and recognition, sonar rangefinding, speed measurement and control, body heat and motion detection, and rechargeable batteries. You can order him fully assembled or in kit format.

A Description of RoPet

The PRC people put in a lot of thought and time to developing this robot. Let's examine just what he is all about.

RoPet is available in two models: RoPet-HR, for consumers, is the complete, fully assembled personal robot, and RoPet-XR, for hobbyists and technical professionals, provides the assembled base unit which has fully exposed, easy-to-reach components. Both models are cartridge-programmable, and three sample cartridges are provided. All soft-

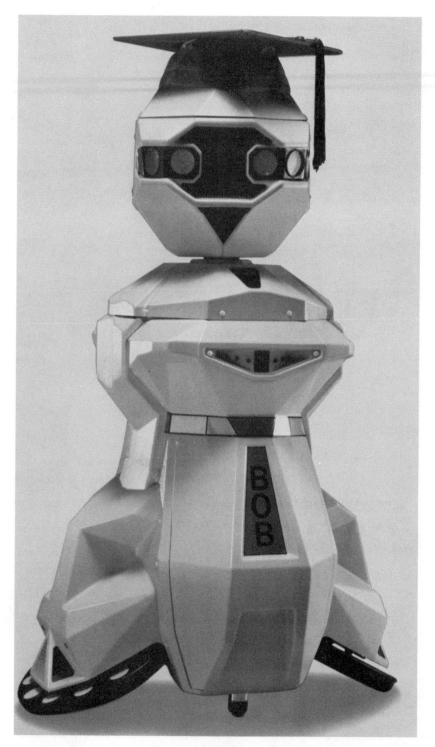

TOPO II is shown with his optional wagon (courtesy Marilyn
Chartrand).

RoPet HR is a fully self-contained personal robot with an on-board Z-80 computer (courtesy Personal Robotics Corporation).

Table 2-6. The RoPet Personal Robot Specifications (courtesy Personal Robotics Corporation).

SPECIFICATIONS:

● COMPUTER

Low Power Z-80 CPU (1 MHz). 8K of 6116 Static RAM is expandable to 56K. 8K ROM provides software for speech recognition, speech and complex sound generation, and motor control. 3 Sample EPROMs provided for demo, home sentry, and RAM initialization. ZIF socket provided for 2716 EPROM driver program. Four blank S-100 slots are available for expansion.

● BATTERIES

Two 12-Volt Gel-Cell Rechargable, 4.5AH each. Low-Power detection. Running time typically 8 hours, recharges overnight on Battery Charger (included). Battery lifetime typically several years.

● POWERTRAIN

Two Independently Driven 4.5" Wheels. Two 20RPM, 14 in.-lb. gearmotors. Two 1" Casters.

● INPUTS

10 tactile position sensors; Polaroid™ Ultra-Sonic Rangefinding; user-definable, on-board speech recognition accurate up to 99+% for any number of 4-word template sets, language independent, speaker dependent; Passive Infra-Red Detector with 35' x 45' adjustable sensitivity; and 2 emittor/detector Infra-Red LED pairs for speed and distance feedback and control of each motor.

● OUTPUTS

Infinite-vocabulary Speech Generation (Votrax SC-01), Complex Sound Generation (TI), and Color Organ (which flashes in synch with Speech).

● PHYSICAL

RoPet-HR: 28" high comes with ivory cover (shown previous page). 30 lbs. (40 lbs. shipping wt.) RoPet-XR: (not shown) is 14" high, has fully exposed electronics (some assembly required). 25 lbs. (35 lbs. shipping wt.) Both Models are 23" wide, include card cage, and have Decahedron base which turns within its own radius. All Materials are ABS plastic.

ware necessary to operate RoPet's on-board features is provided in permanent memory (ROM).

Both models can use the optional Programmer's Package, which allows BASIC programs to be downloaded to RoPet from any terminal or a host computer having an RS-232C standard serial output port. Other options include 56K memory expansion, 1/2 megabyte disc drive, and computerized telecommunications. RoPet-HR is expandable to soon include vacuum cleaner, mechanical arm, and "Butler" serving tray options.

Buying RoPet

Prices for RoPet-HR and RoPet-XR are as follows. Shipping and handling costs are extra.

Fully Assembled and Tested.

- [ ] RoPet-HR: $2,495
- [ ] RoPet-XR: $2,195
- [ ] Motherboard: $1,295
- [ ] RoPet Base (inc. motors, batteries, wheel encoders, etc.): $595
- [ ] S-100 EPROM Burner: $245

Kits.

- [ ] Motherboard (bare) for kits: $145
- [ ] Speech Recognition kit: $295
- [ ] Speech/Complex Sound kit: $195
- [ ] Computer/S-100 Bus Kit: $295
- [ ] Sonar kit: $195
- [ ] Motor-controller kit: $145
- [ ] Infrared Module: $95

To order a RoPet, or for additional information, write or phone the company.

ROBOT SHOP

The "king of personal robot kits,"—that's what some people call Eugene F. Lally, founder and general manager of Robot Shop, an El Toro, California, maker of robot kits, parts, and publications. Lally has also been called other names—like "genius," "in-

novator," and "brain." He was formerly an aerospace engineer and truly an inventor-writer-entrepreneur-businessman extraordinaire. In addition to Robot Shop (before 1983 Robot Shack), the ingenius and prolific scientist owns Dynamic Development Company, which, under his supervision, for 18 years has developed and manufactured a wide variety of consumer, industrial, and aerospace products.

Lally has personally invented over 200 products. Among the products he is responsible for are special clothing, oil additives, watt-saver lightbulbs, a miracle fertilizer, a super-blanket that provides the same warmth as an electric blanket (without electricity!), computer surge protectors, and test-tube pets. The latter product consists of incredible bioengineered critters who come to life in a fishbowl or aquarium within 48 hours after water is added.

Lally's first love is, however, robots. As he told

Eugene Lally, founder of Robot Shop, with the Z-2 robot, built from a kit (courtesy Robot Shop).

me recently when we discussed the field of robotics, "Robots are the most entertaining, educational, fun, and intellectual creations on earth." Lally believes that robots should be available to as many people as possible, and he thinks that his inexpensive and affordable kits could enable this goal. "Robot Shop's kits," he explains, "are designed for kids and adults. They encourage learning and experimentation. Best of all, by building a robot himself (or herself) the individual earns a sense of achievement and personal pride." Lally says you'll get the thrill of your life as you "literally breathe life into your robot."

＼What Can These Robots Do?

Robot Shop and Lally have an exciting group of robots; so let's get down to describing their capabilities. First of all, as I mentioned, these robots are lots of fun, but they have practical and useful applications, too. For example, they can purify air and remove room odors wherever they go in the house. Have the robot go into the kitchen when you are cooking food with objectional odors. Have him follow you into the bathroom. He's great at parties; he keeps the house smelling good and free of smoke. When he is sitting around looking like he is doing nothing, he'll be silently killing bacteria and viruses—he's like a silent servant, always working for you.

The robots can sense smoke and fire and sound an alarm. They sense intruders and warn you of their presence (guard duty). They can drive away pests such as rodents and insects. They can tell you the time of day and remind you of appointments. They can even hear noises and sounds, like the baby crying and someone making noise outside, and tell you about it.

Connect the robot to any television set, and you have a home computer. You can even play computer games against the robot and watch him react with sounds and movements (Robot Emotions) to the game action. As Lally emphasizes, these and many more fun and practical functions will amaze you when you build your own robot. If you can build a hobby model, you can build Robot Shop's robot.

The electrical and electronic subsystems includ-

ed in the kits are completely wired. All you do is attach them to the robot then fasten their input and output wires to terminals with a screwdriver, or in some cases, by soldering with a low-power soldering iron. Robot Shop instructions show you all the steps, yet encourage you to use your own imagination to make your robot "one-of-a-kind." They even explain how you can think up your own special functions for your robot.

The company gives you the option of building robots from complete kits or from parts with which you build and experiment with your own unique robot. Kit designs are based on a building block approach. No matter which kit and options you first purchase, you can always add more options any time, all the way up to computer control.

The computer option provides the ultimate control over your robot. You can easily program the robot with a keyboard mounted right on the outside of the robot. The compact, lightweight Robot Computer Controller is mounted inside the robot and goes everywhere the robot goes. If you already have a home computer, you can interface it with the robot to control its functions.

Emotional Home Robots

Robot Shop has been highly creative in making sure its robots provide first-rate entertainment. For example, you can imagine robots that show *emotions*? Well, Robot Shop has developed a system that allows its robots to display joy, anger, and other emotions. The company says this development represents a breakthrough in robotics technology for hobbyists. The robots now physically react with emotional movements, sounds, and coloration changes to situations to which they are exposed. Humanlike reactions result when home robots do good or bad or when they win or lose at games they play against you. You have heard of artificial intelligence—well here is emotion from machines.

The emotion is accomplished through an option now available for the Robot Shop X-1 robot or for any robot being scratch-built using Robot Shop parts. This option consists of a low-cost computer which mounts in the robot, coupled to electronic devices that impart physical changes to the robot to express emotion. Software is also provided with the option. When you see the robot's physical changes, you recognize his emotional reactions which are appropriate for the situation. "Body language" isn't just for humans anymore.

For example, if you are playing one of the many available games against him like Blackjack or Tic Tac Toe and he wins, his entire body wobbles with the emotion characteristic of a winner, and he lets out a shriek of joy. You can imagine what he does when he loses to you.

The robots can even shiver if the room temperature gets too cool and turn red when their computer is trying to "think" its way out of having walked the robot into a corner of a room. The addition of "emotions and feelings" add a totally new dimension to home robots. The realism and "presence" of the robot are enhanced to the point that home robots almost become a member of the family, like a pet.

Robot Shop also has a unique software package called "Robot Physical Reaction to Game Play." With this system and using the on-board computer, you can play games against the robot and watch him physically react to the mental game action. If the robot beats you at a game, he physically reacts one way; if conversely you beat the robot, he physically reacts a different way.

To obtain this colorful software game cassette software (Part #P-S/W-1), you'll pay $24.95. Not a bad price to add a little personality to your "machine-person" opponent.

Robot Shop Kits

For those who want to obtain a basic exposure to home robots, the DROID-BUG kit is recommended. It is easy to construct and will provide plenty of fun and learning. You can complete it in a couple of hours. It is a complete kit including structure and skin.

If you want to build advanced robots and not spend a lot of money to start with, the Z-1 and Z-2 kits are ideal. All the critical parts are supplied: the motors, electronics, robot controller, and mounting hardware. You provide the structure. Later you can add low-cost options. The instructions tell you everything, including how to build the structure and where to get the structural parts (any local hardware store).

Computer and interface board for robot control. Used for all Robot Shop robot kits, the system comes with full documentation (courtesy Robot Shop).

For those who really want to get into the hobby of robotics in a big way, there's the X-1 robot kit. It comes complete with structure and skin members. You may also select some of the exciting optional functions available for the X-1 to enhance its capabilities and for added fun and learning. Construction of this kit is also simplified but takes more time to complete than the DROID-BUG kit.

For those with great imagination and more skill, Robot Shop will provide all the robot parts, sensors, and publications that you will need to design, construct, and experiment with your own robot creation. This is a great way to exercise your creativity, but don't expect to put a robot together in a few hours.

You can mount the Robot Shop on-board robot computer (about $200) on your robot and have a high degree of control by programming the robot to do what you want. Also, interfacing the robot to any home computer is possible.

Each kit provides step-by-step construction guidelines for those who want to build a robot according to the plans. You can, however, change the construction details and also add to it using an optional list of parts, sensors, and functions. Rearrange the design any way you want and fashion a robot using your own ideas. As the people from Robot Shop point out, this is part of the fun of the personal robot hobby—creating a unique "one-of-a-kind" robot. They say you can take their basic suggestions and build and elaborate upon them with your own concepts and ideas to come up with something new and amazing. If you choose to provide your robot with a wooden body, you'll no doubt learn something about basic carpentry. If metal is chosen, that will provide you with quite different skills.

DROID-BUG. DROID-BUG is a lovable and playful robot that is completely self-contained and easily assembled. It is great as a first robot project or as a companion to more mature, big-brother robots.

The tiny bug runs about the floor or ground and senses obstacles it touches with a feeler. Then, it turns away and continues in a new direction until it meets another obstacle. The robot makes a loud buzz sound like a hurt bug when it senses something in its path.

DROID-BUG acts alive as it scoots about, frantically avoiding anything that trys to get in its way. This robot kit will provide lots of fun and teaches basic robot construction. The cost is $129.95 plus shipping.

Its dimensions are: 9 inches in diameter, 7 inches tall, and dome-shaped. It weighs 4 pounds. Included in the kit are:

- ☐ Body, frame, and dome structure
- ☐ Dual motor drive with gear reducer
- ☐ Rubber drive wheels
- ☐ Obstacle feeler sensor
- ☐ Noisemaker—bug sound
- ☐ Skin features/markings
- ☐ Wire and connectors
- ☐ Assembly hardware
- ☐ Assembly instructions
- ☐ Battery holder for four C batteries.

Tools needed to assemble: common hand tools, a low-power soldering iron, and spray paint. An option for DROID-Bug is a nickel cadmium battery set and charger to replace flashlight batteries for longer robot on-time and for savings in battery costs.

The Z-1 Robot. The Z-1 robot is more ad-

DROID-BUG is a self-contained robot that scoots energetically about, avoiding obstacles and making buzzing noises when frustrated (courtesy Robot Shop).

vanced than DROID-BUG. It, the Z-2, and the X-1 are designated with letters and numbers only so you can have the pleasure of dubbing the robot with your own special name or title. The Z-1 kit includes motors, electronics, robot controller, and mounting hardware to make a home robot of up to 25 pounds total weight. Everything you need to build a robot is included, except for the structure and skin, which you provide; the instructions tell you how and where to find materials (any local hardware store). Cost of the Z-1 kit is $149.95 plus shipping. The Battery pack ($79.95) is extra.

Included in the Z-1 robot kit are:

- [ ] Complete instructions (includes how to build and what to use for structure and skin)
- [ ] Dual motors
- [ ] Wheels
- [ ] Robot controller (controls up to 6 robot functions at one time with on-board switches and timing circuit; the brain of the robot)
- [ ] Third wheel caster
- [ ] Hardware for mounting the motors, robot, controller, caster, and terminal strips.
- [ ] 2 rolls of hook-up wire
- [ ] 2 terminal strips

All of the many options I will soon review are compatible with the Z-1 and Z-2 robots. So, you can first build the basic robot then later add accessories—senses, smell, speech, and so on. Also, a battery set is offered as an option.

The Z-2 Robot. The Z-2 robot is the most rugged of kits offered by Robot Shop. This is a 100-pound class robot similar to R2D2 from *Star Wars*. An example of a Z-2 robot was the "machine person" pictured earlier in this section with Robot Shop president Eugene Lally. The Z-2 kit includes motors, electronics, robot controller, and mounting hardware—everything you need to build an advanced large robot except for the structure and skin, which you provide. The instructions tell you how

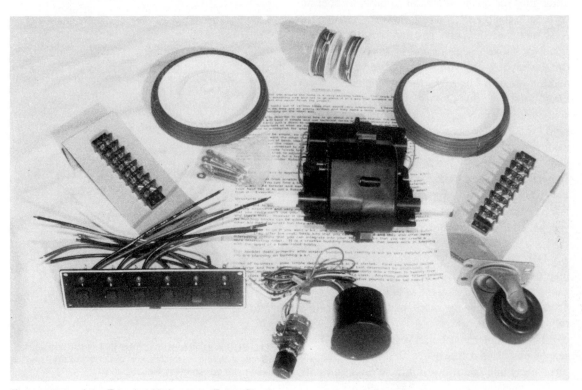

The contents of the Z-1 robot kit (courtesy Robot Shop).

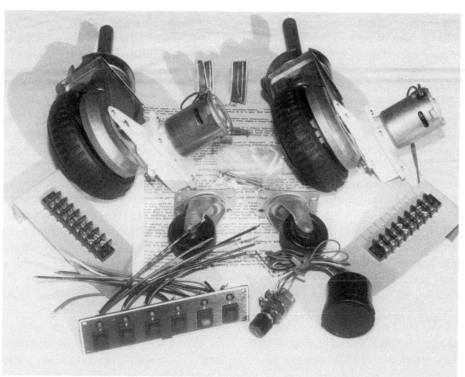

The Z-2 robot kit which enables a hobbyist to build a sturdy, 100-pound robot (courtesy Robot Shop).

and where to find materials. Cost of the Z-2 kit is $249.95 plus shipping. The battery pack is extra ($89.95).

Included in the Z-2 robot kit are:

- ☐ Complete instructions (includes how to build and what to use for structure and skin)
- ☐ Dual motors
- ☐ Wheels
- ☐ Robot controller (controls up to 6 robot functions at one time with on-board switches and timing circuit, the brains of the robot)
- ☐ Third and fourth wheel casters
- ☐ Hardware for mounting the motors, robot controller, casters, and terminal strips
- ☐ 2 rolls of hook-up wire
- ☐ 2 terminal strips

The X-1 Robot. Unlike the other kits, the X-1 kits comes with everything needed, including the structure, to construct an economical personal ro-

bot with many functions and features. This is Robot Shop's top-of-the-line kit. Cost of the kit is $399.95 plus shipping. The battery pack is optional and costs $79.95.

Functions of the X-1 robot include:

- ☐ Locomotion—complete ground mobility
- ☐ Complete ground mobility control
- ☐ Controlled by on-board control panel
- ☐ Primary robot functions are selected and cycled ON/OFF with a timer adjustment
- ☐ Sounds/noises
- ☐ Lights his heart/eyes

Dimensions are 22 inches long, 12 inches wide, and 18 inches tall. It weighs 25 pounds (with all options installed).

Included in the X-1 robot kit are:

- ☐ Body, frame, and skin panels
- ☐ Dual motor drive and gear reducer

- ☐ Rubber drive wheels
- ☐ On-board control panel
- ☐ Function cycle timer
- ☐ Noisemaker
- ☐ Lighted robot heart/eyes
- ☐ Body and skin features/markings
- ☐ Wire and connectors
- ☐ Assembly hardware
- ☐ Assembly instructions

Tools needed to assemble include common hand tools, a low-power soldering iron, and spray paint.

Options

If the descriptions of Robot Shop's kits excite you, check out the options you can buy to upgrade your robot. Among these options are the following:

Hearing Sense. Turns on and off any selected robot function(s) when it "hears" your voice, hand clap, or whistle. The first sound switches any function(s) on until it "hears" a second command sound, which turns the function(s) off. It comes complete with mounting hardware and instructions. Cost is $29.95 plus shipping.

Human Intrusion Detector/Alarm—Ultrasonic Sensor. This is the guard duty function with a range of 15 feet. It lets the robot sense unfriendly humans and animals as they approach and protect itself or the inside of your home. Sensitivity adjustment is included and 12 Vdc. When a human is detected, it switches on any robot function, such as noisemaker, loud chirping horn alarm, or any motion function, such as robot spin around. It comes complete with mounting hardware and instructions. The cost is $89.95 plus shipping.

Human Intrusion Detector/Alarm—Infrared Sensor—Guard Duty Function. Range is 35 feet. It is the latest advancement in passive infrared technology and provides detection of humans and animals approaching the robot. The area protected includes 35 feet long × 20 feet wide forward of the robot. Sensitivity adjustment is included. Cost is $169.95 plus shipping.

Air Purifier. It silently and electronically cleans household air, removing odors, smoke, fumes, pollen, dust, bacteria, and viruses. Wherev-

With the X-1 kit, you can build a personal robot with a number of fascinating functions (courtesy Robot Shop).

er the robot goes in the house, he does all this electronically with no expensive replacement filters needed. A test probe is also included to prove performance. Cost is $99.95 plus shipping.

Smoke and Fire Sensor. This option is a Dual-chamber, ionization-type sensor. It is ultraresponsive in the detection of products of combustion. When triggered, the robot instantly sounds a shrill-piercing alarm. Locate your robot in the hallway near bedrooms for nighttime protection. Take the robot on family trips for protection in hotel rooms and in camp tents, campers and RVs. It has a built-in test switch. Cost is $19.95 plus shipping.

Pest Controller. Here is the scientific way to repel pests such as rats, mice, roaches, flies, fleas, mosquitos, bedbugs, spiders, ticks, crickets, moths, squirrels, skunks, etc. The robot will electronically drive away these pests with powerful blasts of ultrasonic sound waves which are imperceptible to people and pets. Cost is $99.95 plus shipping.

Water Gun Function. This option squirts water out of the robot like a water pistol. Use with on-board controller or connect to Human Intrusion Detector/Alarm. It uses 12Vdc. Cost is $39.95 plus shipping.

Swinging Arms Kit. This option is for use on tall robots of the 50- to 100-pound class. It includes two 12 Vdc motors with 90-degree motion and return to create the normal swing of arms, two styrofoam arm kits, mounting hardware, wiring, and instructions. Weight is 6 pounds. Cost is $149.95 plus shipping.

Balloon Inflation Function. This option permits the robot to blow up balloons—great for parties—also to inflate automobile and bicycle tires and vinyl swimming pools. It is called the *Belly Button* option because a fitting protrudes from the robot like a belly button to which you connect balloons etc. to be inflated. It includes an air compressor, hose, mounting hardware, and instructions, with 12 Vdc. Weight is 5 pounds. For use on robots of the 50- to 100-pound class, it costs $49.95 plus shipping.

Night Light. A light assembly mounted inside the robot shines through the skin. Side panels of the robot appear to glow at night, and the robot can be placed on a table or on the floor where it will act as a night light. It comes complete with mounting hardware and instructions, with 12 Vdc. Cost is $14.95 plus shipping.

Ultrasonic Ranging System. This system can sense and measure distances of objects out to 35 feet. It is used to warn the robot of obstacles in its path. It can be set up to turn the robot away from obstacles at any preset distance. This can be the robot's "seeing" sense. The robot can be used as an "electronic tape measure" and measure dimensions of a room, etc. Cost is $299.95 plus shipping.

2-Channel Remote Radio Control System. This option controls the forward motion of travel and steers the robot. It will permit you to make the robot travel anywhere you want remotely. Range is 100 feet. It includes a two-channel, handheld transmitter, a two-channel, robot-mounted receiver; mounting hardware, and instructions. Cost is $349.95 plus shipping.

Robot Computer and 8 Function Control Board. This board mounts on-board the robot. It includes a 2K RAM Timex/Sinclair 1000 computer, membrane keyboard, interface board with eight channels of relay controlled outputs, interconnecting cables, mounting hardware, installation, and operating and programming instructions. The computer controls eight robot functions in any sequence and for any duration you program. This is the most versatile way to control your robot. It can also be used as a home computer by connecting the robot to any television set. Game software is also available, and you can play against your robot. Cost is $249.95 plus shipping.

And There's More

The Robot Shop offers several other services and products for robot enthusiasts, including membership in the Robot Shop Club. Membership cost is $10, which entitles you to receive periodic *Personal Robot* news bulletins and announcements of new robot parts and functions.

For buyers of its robot kits, the Robot Shop offers a free answering service for questions. Just send any questions you have on a separate sheet of paper. Leave some blank space after each question because they will write directly on your paper. Send it along with a self-addressed stamped envelope. Include this with your initial or subsequent orders or any time later while you are assembling one of the kits.

Robot Shop also offers three publications: a catalog and two illustrated booklets:

☐ Catalog of Home Robot Kits and Parts ($3.00)
☐ Designing Personal Robots—Basic Ideas ($9.95)
☐ Complete Handbook of Home Robots— Basic Through Advanced ($15.00)

To buy a Robot Shop kit or other product, or for more information, write to Robot Shop.

TM

SERVITRON ROBOTS

Now for something completely different . . . well, almost. Got a hankering for your very own chef, butler, or French maid? Then, Servitron Robots of Denver, Colorado, has just the thing for you: inflatable servant robots costing an affordable $100 or less.

Servitron Robot offers all the household help you'll ever need: a French maid, a butler, and a master chef (courtesy Servitron Robots, Inc.).

Servitron is the brainchild of two high-tech entrepreneurs, Gary Schlatter and Bert Matissen. Sensing that the great American middle class would love to be able to afford its own live-in household help and realizing the popular trend toward robots, the two partners created the Servitron robots. Lightweight and easy to assemble, these remote-controlled, battery-operated creations have caught the market by story. The media have pictured the robots in a number of poses, and many major department stores are now stocking the servant robots, including Joske's stores in Texas and Jordan Marsh stores in Massachusetts.

Servitron really had a coup when a major magazine hailed its robots as the gift of the year. Then the upscale Neiman-Marcus Department Store marketers in Dallas decided to offer a Servitron in its fabulous Christmas catalog, dressed as Santa Claus. As a consequence, the robots are selling like wildfire—about $4 million so far.

Servitron's Bert Matissen says you can dress the robots up any way you want. "You should see the way *Playboy* magazine at first conceived dressing and displaying the French maid," he recently quipped. Each comes dressed in one of the three basic uniforms, however, and they must be ordered individually as "French Maid," "Butler," or "Master Chef."

Features

These fascinating robots operate at distances of 75 feet and more. A simple handheld control transmitter makes the robots turn, go forward, or reverse. The motor for the robot platform is driven by four D-cell batteries, and the transmitter is powered by a single A-volt battery.

More features include:

☐ Stands just over 4 feet tall.
☐ Comes complete with uniform and aluminum serving tray.
☐ Traction tread wheels. Runs on pile carpets and hard floors.

You can have these home helpers up and run-

ning in just a few minutes. As Servitron points out, "Better help is hard to find," and though these robots are not intelligent, computer-controlled models, they are certainly perfect as party attendants. And what conversation pieces!

Buying a Servitron Servant

If you want your next party to be a smash or if, say, breakfast in bed is your style, you can order your robot servant from Servitron.

IOWA PRECISION ROBOTICS

Take three self-confessed computer nerds, several sheets of heavy plastic, sturdy steel plate, a Motorola 68000 microprocessor, sonar sensors, a room full of Servo motors, and electric wires and what do you get? Why, a 150-pound robot, of course. MARVIN MARK I, his manufacturers claim, is the most sophisticated personal robot ever produced.

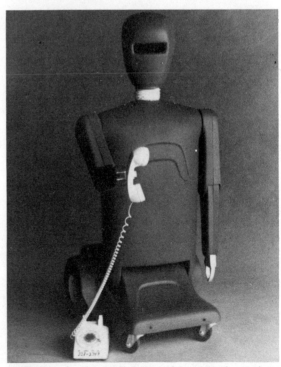

MARVIN costs a lot, but he's powerful and tough, a unique combination of brawns and brain (courtesy Iowa Precision Robotics).

MARVIN is designed as a teaching aid for industry, technical schools, universities, and advanced hobbyists.

MARVIN I moves around the room, talks with a synthesized voice using his 500-word vocabulary, moves his head, has sonar ranging on board, and has two 6-axes arms that can be programmed to work simultaneously. This latter feature is practically unparalleled among personal robots.

The robot can be operated as a totally independent personal/business computer with its on-board, 3 1/2-inch, high-density floppy disk drive. Just plug in a terminal, and you have a personal computer with more power than most business computers. For example, the computer is more powerful than an IBM PC and equal to the Apple Lisa. MARVIN comes standard with 128K RAM on board which can be expanded to 512K. With his 6-slot S-100 bus, you can expand his memory beyond 512K or add your choice of the S-100 cards on the market. MARVIN's computer brain is a Motorola 68000 CPU microcomputer with a 16-bit data path and a 24-bit address bus. The 68000 is part of a new generation of ultrahigh-speed (state-of-the-art) microcomputers.

The MARVIN MARK I has a humanoid torso with a shatterproof, vacuum-molded body skin. Under the skin is an aircraft aluminum chassis containing high-quality machined parts to give the MARK I a long life span. The MARK I can be operated without the outer body skin to demonstrate his mechanics.

Dan Knoblaugh, president of Iowa Precision Robotics, MARVIN's manufacturer, says that "to train students and employees for a career in robotics you need a challenging, sophisticated, open ended yet forgiving teaching tool which you can expand and update for future challenges. MARVIN MARK I fits that description to a T."

The Company

Iowa Precision Robotics was founded by three Milford, Iowa, enterpreneurs—Dave Gossman, Matt Plagman, and Dan Knoblaugh. I talked with Dan at Albuquerque's First International Personal Robot Congress in 1984 and was impressed with his enthusiasm for personal and educational robots. At that time, MARVIN was still in the research-and-development stage (a mock-up was on display), but Dave Gossman, chief executive officer at the Iowa firm, told me that production began in earnest in the summer of 1984, adding that orders were pouring in from around the world.

The Robot

If this robot can do what it's makers say it can do, it's a breakthrough for the industry. Certainly, the product specifications are impressive. MARVIN is reputed to have a hand dexterous enough to hold a cigarette. He can move his head, can easily bend over and pick up a tool from a table, and can talk, all at the same time. His blue plastic body is tough and should hold up to rough treatment over time.

Incidentally, MARVIN stands for *M*obile *A*nthropomorphic *R*obot. The last three letters, VIN? Knoblaugh jokingly confided to me that this is a "VINtage high tech robot," meaning it's fully state-of-the-art.

MARVIN appears to be an ideal robot for high schools, businesses, technical schools, and colleges which offer robotic courses for employees and students. His cost ($5,995) is not prohibitive, considering the advantages he has to offer as a mobile training aid with his two flexible arms and grippers and powerful microcomputer brain. Table 2-7 shows cost and specifications for MARVIN MARK I.

ROBOTLAND

Remember about 6 or 7 years ago when they laughed at people who dreamed of a chain of computer stores stretching across the United States? Today, several such entrepreneurs are rich—very rich—because chains like ComputerLand now list dozens, even hundreds of stores in their annual reports.

George Cretecos has a similar dream, though with a different product, and naturally some may call his ideas a little far-out. Don't bet against George, though. What he has in mind is a chain of hundreds of robot retail stores, and in Boca Raton, Florida, in the middle of Silicon Beach, George and a few partners have already opened their first store.

George says that "Soon, they'll be opening a Robot Shack and a Robot Hut down the street" from

Table 2-7. MARVIN Price List and Specifications.

MARVIN MARK I robot $5,995

(With standard 128K and one disk drive)

MARVIN PRICE LIST

The standard 128K bytes of memory in MARVIN should prove more than adequate for all programs. For very advanced work you might wish to extend the memory.

On-board memory factory extended from the standard 128K bytes:

Pick one of	Up to 256K bytes	$310
three levels	Up to 384K bytes	$595
	Up to 512K bytes	$875

Second disk drive factory installed	$ 245
Diskettes	$ 7
10 for	$ 67
Non-spill deep discharge battery	$ 90
(Two needed for MARVIN)	
Additional deep cycle battery chargers	$ 85
(One comes with MARVIN)	
CP/M 68K operating system	$ 350
(By Digital Research)	

Other Software Priced as Available.

MARVIN MARK I User Manual only .... $ 40
(Manual free with robot, price refunded if robot purchased later.)

Note: Buyer must supply two batteries and standard RS-232 terminal for operation. Additional chargers and batteries are needed only for extended mobile use.

MARVIN newsletter .... FREE

GENERAL SPECIFICATIONS

Computer system:
> Model 68-100 Industrial Robot Controller
>> By Iowa Precision Robotics, Ltd.
>> Entire system battery powered.
>> Power system provides blackout-proof operation.
>> Onboard S-100 mainframe.

Processor:
> Motorola @ 68000 bit CPU on single board computer.
>> A. Direct addressing of up to 16 Mbytes, 8 MHz system clock.
>> B. ROM monitor.
>> C. Can be used as a stand-alone computer.

System bus:
> 8-slot S-100/IEE-696 Standard Bus.

Memory:
>> A. Standard 128K bytes RAM, on motherboard.
>> B. Sockets on motherboard allow expansion to 512K bytes RAM.
>> C. Can be extended to 16 Mbytes with plug-in S-100 cards.

ROM:
> Boot ROM by Iowa Precision Robotics, Ltd.
>> A. Standard monitor ROM is 1K byte.
>> B. ROM can be expanded by adding cards on the S-100 bus.

I/O ports:
   A. Control inputs supported by any standard RS-232 terminal.
   B. Two full RS-232 ports.
   C. One 8-bit parallel port, with full handshaking.
   D. System I/O ports on motherboard, used for communications with control terminal.
   E. Separate and direct link between the computer and the servos.
   F. I/O ports can be expanded with S-100 cards.
   G. I/O ports allow MARVIN MARK I to by synchronized with surrounding equipment.

Disk Drive:
   Ports or plug-in cards.
      A. 3 1/2 inch single sided microfloppy, SA300, made by Shugart.
      B. Stores 390 Kbytes formatted per drive.
      C. One disk drive standard, expandable to two. Controller can link to external 5 1/4 inch floppy disk drive.
      D. The disk drive controller is a WD1770 by Western Digital. It will control up to 4 drives, 3 1/2 or 5 1/4 inches.
   Languages supported:
      A. ERPL (Educational Robot Programming Language).
      B. Any language supported by the 68000.

Operating system:
   A. ERPL.
   B. CP/M 68K by Digital Research Inc.
   C. Other user-supplied 68000 operating system.

Software provided:
   ERPL programming language, full test and demonstration package, diagnostic software, text-to-speech program, servo control package, and ultrasonic ranging software.
Software limitations:
   The amount of program steps that can be stored is limited to memory and bulk storage capacity.
Servo control system:
   A. Custom S-100 card with 16 full four-quadrant power, MOSFET pulse-width modulated servo controllers.
   B. Pulse-width modulation is under software control.
   C. Card has format error checking PROM.
   D. User may install up to four cards for 16 more servo channels per card.

Voice card:
   A. Custom S-100 type, full address decoding card with SSI 263 speech chip.
   B. Text-to-speech algorithm in software.
   C. Unlimited vocabulary.
   D. 64 phonemes.
   E. Voice recognition option:
      A. State-of-the-art VLSI voice recognition chip set.
      B. Unlimited amount of recognizable words on disks, with a menu of 128 words at a time.
      C. 98% recognition accuracy.
      D. User-specific.

Sensors:
   A. Ultrasonic ranging system operates up to 25 feet.
   B. Based on the polaroid ranging system.
   C. Additional sensors could be supported by user-supplied S-100 I/O cards.

Power system:
   A. Battery: Two 55 A/H Deep Cycle--one for electronics, one for motors.
   B. dc to dc supply for computer system to conserve battery power.
   C. Quick-change battery system.
   D. System has low voltage detectors and reverse voltage protection circuits.
   E. Onboard battery charger, 110 V.A.C., 60 Hertz, 5A.
   F. Run-time on motor battery is 3 to 4 hours, dependent on amount of motor usage.
   G. Run-time on electronics battery is approximately 10 to 11 hours.

Arms:
  A. MARVIN MARK I has two 6-axis arms capable of lifting up to five pounds each.
  B. Shoulder: 0° (down)-135° (up)--a "bowling" motion; and 0° (down)-90° (sideways)--a "jumping jack" motion.
  C. Upper arm rotates 0°270°-an "arm wrestling" motion.
  D. Elbow: 0°90° in "plane" of arm; plane set by upper arm rotation.
  E. Wrist: 360° rotation.
  F. Hand grip range: 0-4 inches.

Waist:
  A. Range of 0° (upright) to 45° (forward).
  B. Allows MARVIN MARK I to reach the floor when bent forward.

Neck:
  A. Rotates around 360°.
  B. Tilts forward 30° and back 15°.

Axis motion:
  A. Most axes travel through 45° per second.
  B. Typical limit-to-limit transit time is 3 seconds.

Mobility base:
  A. Speed of up to 25 inches per second.
  B. Can climb or descend a 10° ramp.
  C. Can cross-traverse a 10° slope.
  D. Ten-inch pneumatic tires give enhanced traction and quiet operation with no scuff marks.

Body and Frame:
  A. Anthropomorphic body shell.
  B. Vacuum-molded KYDEX-100 body in Cadet Blue.
  C. Aluminum Framing.
  D. Total weight of robot is approximately 150 pounds.
  E. Dimensions: 4'3" tall, 25" wide, 28" deep.

Support offered:
  Newsletter, User Group, Software Bank.
Manuals supplied:
  User's Manual, ERPL Software Manual.
  Robotics education course is under development.

his store. By then, however, he hopes to have a big head start in the race to become the world's top robot retailer. Some would say George's plans are grandiose. That, after all, is what the experts said about Stephen Jobs and Steve Wozniak in the mid-70s when they began their drive to launch Apple Computer, Inc.

The ELAMI Robots

Robotland retails most major models of robots, including MARVIN, RB5X, and HERO I. It also markets its own models, called ELAMI. ELAMIs come in three versions, each of a different size and capability. All feature on-board computers, motor-

ized arms, and a brightly colored body with facial expressions. A description of each robot and specifications follow.

12-Inch ELAMI. 8-bit microcomputer, custom design, 2K of ROM and 128K RAM memory for storage and operation of programs. It features a two-color LCD face, four facial expressions, rotating head, LED light display in head, two motorized moving arms and grippers, speech synthesis, and 24-key miniature computer keyboard surrounded by 17 LED light display. The unit is programmable up to 64 minutes of continuous operation. The robot walks straight even on thick carpet, can make exact 180-degree turns, can retrace its pathway, can move forward, back, left turn, right

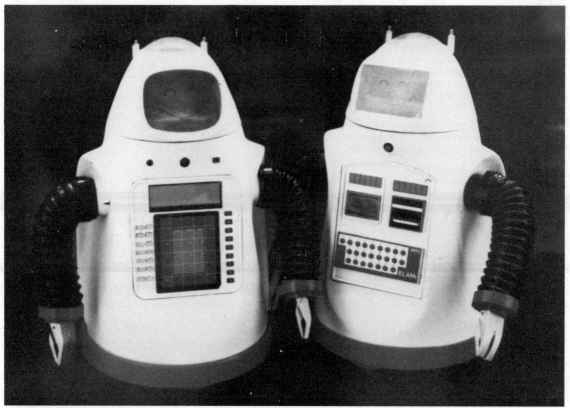

The ELAMI robot, offered exclusively in the United States and Canada by Robotland, comes in three sizes: 12-inch, 18-inch, and 36-inch. Each has an on-board microcomputer (courtesy Robotland, Inc.).

turn, and operate left wheel or right wheel with all functions in two speeds. It is equipped with a preprogrammed demonstration showing all features. ELAMI comes equipped with individualized computer I.D. card with serial number and security code. When bumped, two tactile sensors are activated, and the robot responds by turning 90 degrees to correct its pathway. One infrared sensor allows the robot to detect and avoid obstacles. The robot is operated by one 9-volt and four C size batteries.

18-Inch ELAMI. 8-bit microcomputer, custom design, 4K ROM and 4K RAM memory expansion. ROM/RAM cartridge for memory expansion including talking calculator, speak and spell, speak mathematics, talking babysitter, maze solving, room mapping, and voice recognition. It features a two-color LCD face with four facial ex-

pressions. The simulated lips are synchronized to speech synthesis. The two motorized arms with grippers have three axes of movement. The front panel includes nine function keys and an LCD display for data inputs. A touch keyboard and 14 command keys, infrared sensors, LED light display, and a function switch, all-clear switch, and set-time switch are on board the robot. The back panel includes a speaker, infrared sensors, software cartridge slot, and on/off switch. The driving motors include three forward and backward speeds. The robot is capable of 180-degree exact turns. The unit is equipped with a talking clock with six to seven fixed alarm messages and one programmable alarm message. It is also equipped with tactile sensors around the base of the unit as well as an RS-232C interface port. The 18-inch model also comes with security code and optional remote control joystick.

The robot's head will rotate 90 degrees in either direction.

36-Inch ELAMI. 64K microcomputer with full-sized keyboard, data recorder, and one disc drive (second available as an option). It features a nine-inch color monitor with synchronized facial expressions. The head will rotate 45 degrees in either direction. The dome of the head is equipped with a plasma light display. The robot has two articulating arms with five axes of movement for each. Each arm is capable of lifting a weight of 25 pounds and is computer- or radio-controlled. The system is equipped with ultrasonic range finders and infrared sensors, as well as tactile bumper sensors. The system can operate by other computers via the RS-232C port or by its own on-board computer or radio control. Standard speech includes a 2,000-word vocabulary. Voice recognition, upgradeable to 263 words, is available. Motors include two 12-volt dc units with three forward and backward speeds and variable speeds of 6 to 24 inches per second. Batteries include one 12-volt, 24-amp hour and two 12-volt, 15-amp hour rechargable units.

Buying the ELAMI Robots

Barring a visit to the Robotland store in Florida, you'll have to acquire one of the three ELAMI robots by mail order from the company.

RHINO ROBOTS

Rhino Robots, a Champaign, Illinois firm, is no newcomer to the world of robotics. The company has earned a well-deserved reputation for building quality robot education systems. Its XR robot training programs are in use at dozens of colleges, technical schools, and other training centers in the United States and Canada. Professional educators continually heap praise on Rhino's systems.

Successful with its tough and useful educational arm robots, Rhino set out in late 1983 to create a personal robot vehicle that may appeal to a broader mass market, yet retain excellent educational val-ue. The result was the SCORPION.

Your Very Own Interplanetary Explorer Vehicle

How would you like to own a miniature Lunar Reconnaissance Vehicle (LRV) for exploring the planet earth? Rhino has developed the SCORPION, a floor mobile robot that looks like the moon vehicle and is as exciting for earthlings as the LRV was for the astronauts. The SCORPION is a sophisticated device that will appeal to serious computer hobbyists of all ages and can be run from any host computer with an RS-232C interface.

In its 11-x-14-inch body is a 6502 microprocessor and two 6522 interface chips which provide control over two ports and 32 I/O lines. It has four programmable timers, two of which can be event counters.

Like the Lunar Rover, the SCORPION has many channels for two-way communication and therefore is ideally suited for a wide range of artificial intelligence experiments. With its two-axis optical scanner that covers a 300-degree range in the vertical and horizontal plane, the SCORPION can project objects of varying degrees of brightness on the CRT of the host computer. The optical scanner looks like the umbrellalike receiving dish on the high-gain antenna that projected over the Lunar Rover. In addition, two ground trackers can read a shiny tape on the floor and follow a designated path over flat ground.

SCORPION is highly sensitive to bumps on its sides. Eight microswitches can report collisions on all sides and indicate whether they occurred on the front or on the back. A speaker can supply appropriate groans (or squeak with delight) when a collision occurs. The programmer can select the frequency and duration of sounds (several Rhino staff have been writing SCORPION music in their spare time). While it sings to you, its two front eyes can be programmed to open and close independently and give a friendly wink. The wheels of the SCORPION are driven by two independently programmed motors and move in either direction at any one of 99 different speeds.

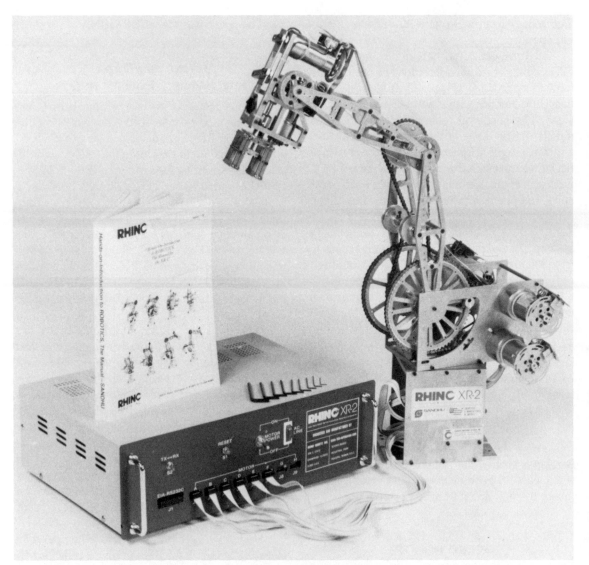

This XR-2 training system offered by Rhino Robots is highly recommended by engineers and educators. Many accessories are available. The cost is about $2,850 (courtesy Rhino Robots).

Basically, the SCORPION controls the following:

- [ ] 4 Motors
- [ ] 1 Speaker
- [ ] 2 Eyes
- [ ] 2 Ground scanners
- [ ] 8 Switches

Optional items include:

- [ ] 2 more motors
- [ ] 16 additional I/O lines

The microprocessor on board has 8K of EPROM and 2K of RAM and is fully expandable to a 64K system. The SCORPION's power requirements are minimal. It needs 12Vdc to run and is satisfied with a 12-volt battery.

The new RHINO SCORPION is especially designed for the serious computer hobbyist. With its 11-x-14-inch carriage and rotating optical scanner, it resembles a lunar explorer. It is completely software programmable with 32 lines of I/O. The computer controls four motors, one speaker, two eyes, two ground tracks, and a two-axis optical scanner (courtesy Rhino Robots).

Having Fun with SCORPION

The SCORPION can be persuaded to carry out several mundane tasks even though its major talent lies in artificial intelligene (AI) experiments and in running SCORPION races. For instance, since the holes in its body are compatible with erector sets and meccano parts, the user can build a traylike contraption on its back. The tray can be loaded with the evening paper, food, and whatever else a person returning home at night might desire. The SCORPION can be programmed to follow a predetermined path to meet its owner when the optical scanner picks up the light that is switched on to signal the return home. SCORPION could even greet the arrival with a theme song and a wink. The SCORPION is equally adept at following a flashlight or a piece of foil on the back of the person it is trailing. Just think of its potential for practical jokes!

Like the Lunar Rover, the SCORPION will require some assembly before going into action, but don't worry, a 250-page manual with many diagrams comes with the package. The Lunar Rover took 17 months to put together, but the SCORPION is estimated to take only a day or two. Of course, that doesn't count the time you will be taking with its appearance. Probably you'll first want to decide what personality you want for your SCORPION. Should it be fierce, weird, or elegant? Then you will select and apply the proper paints, decals, designs, and whatever else your plans require.

The SCORPION will be an endless source of fun and challenge for a computer buff or engineer and an excellent teaching/learning device for schools, parents, or children. It will make computer games look tame and give you a head start on the revolution in artificial intelligence.

Buying SCORPION

The suggested retail price of Rhino's SCORPION is $660. Periodically, the company offers a special or discount price; so you may wish to inquire about that. To order either SCORPION or XR-2, or to obtain additional information, phone or write Rhino Robots, Inc.

FEEDBACK

Buy a robot from Feedback and you're dealing with a leader in meeting the needs of technical education for over 25 years. Feedback is known for excellent equipment covering the full spectrum of high-tech processes, including communications, microprocessors and microcomputers, robotics, electrical machines, instrumentation and control, electricity and electronics, and hydraulics and pneumatics. All equipment has been designed to stimulate and motivate students on the vocational-technical level by transforming textbook theory into exciting and enjoyable hands-on experience. Geared for both teaching and learning flexibility, the equipment helps instill a broad base of knowledge, an understanding of the basic processes, and the desire to go on to further education.

Feedback has a number of nonmobile, arm teaching robots. Most are exclusively for educational, industrial, and military uses; however, one, the ARMOVER, seems to me a useful educational tool for the home learner as well. In addition, Feedback offers a low-cost, turtlelike robot called, appropriately, the ARMADILLO. So, let's examine these two worthy robots.

The ARMOVER Teaching Robot

The ARMOVER SSA 1040 is an extremely versatile and affordable instructional robotic arm designed for microcomputer interface. The ARMOVER simulates actual production situations for a fraction of the cost of larger industrial robots.

The unit incorporates several features from industrial robots, including a rotational waist, 5-axis revolute arm, and two-fingered gripper, which can sense both the presence and size of an object as the grip closes.

Through direct interface with an Apple II, TRS-80™, or Commodore computer (with minor modifications), the robot can offer a programming capacity. Electromechanical sensors fitted to all axes allow the robot to be automatically initialized to a predetermined position under software control, eliminating the restrictive manual setup requirements of open-loop operation. Free multiplexed input channels are provided for connection of external sensors.

The five-jointed arm has a lifting capacity of 16 ounces fully extended, and a speed range from 2 to 6 inches per second. The two-fingered gripper can open up to 3 inches, exert a force up to 3 pounds, and judge the size of an object within 1/16 of an inch. Cable control from stepper motors in the body enables the robot to move more quickly while providing greater lifting power.

The ARMOVER SSA 1040 comes complete with instruction manual and Armsoft instructional control program on diskette. Price is $2,210 plus the cost of the interface package.

The ARMADILLO

Aptly named, the ARMADILLO (Model EMR 1020) computer-controlled, educational robot is ideally suited to teach robotics principles. Under computer control, the ARMADILLO will run around forward, backward, and to the left or right at a speed of 15 feet per minute. Each wheel is independently controlled. Whenever the robot encounters an unmovable object, touch sensors send data back to the remote computer, which then directs either evasive or exploratory action. ARMADILLO has blinking "eyes," beeps in either of two tones, and when directed by the computer, will press down a pen and chart its programmed progress on paper.

According to the company, the ARMADILLO connects to the input/output ports of the ZX81

The ARMOVER is an extremely versatile instructional robotic arm designed for control by an external computer (courtesy Feedback, Inc.).

The ARMADILLO teaches robotics principles (courtesy Feedback, Inc.).

(Timex/Sinclair 1000), the AIM65, or other microcomputers. The interface circuit enables the robot to be treated as a memory-mapped I/O device so data can be sent to and received from the robot as if it were another memory location in RAM or ROM.

The robot comes fully assembled and tested or, for increasing learning, is available in kit form for self-assembly. It runs on a dc supply of 9 to 15 volts drawn from the host computer. Price of the fully assembled model is $525.

How to Buy Feedback Robots
To order a Feedback robot or catalog, or for additional information, contact the company.

STOCK MODEL PARTS
Fun and education at a low cost; that's what Stock Model Parts offers buyers of its two robots: the LINE TRACER II and the MEMOCON CRAWLER. Stock Model Parts is a division of Designatronics, a New Hyde Park, New York, company that is a major supplier of electronic and mechanical parts. The company is known in the parts industry as a very reliable firm. Let's take a look at their small, but worthwhile, robots. Both are offered in easily assembled kit form.

LINE TRACER II
The LINE TRACER robot utilizes an infrared light sensor to guide it along any drawn black line pathway. A black felt tipped marker or tape can be used to create a straight or curved pathway, at least 3/8-inch wide, on any white paper or white floor area. Applications include school science projects and robotics courses to learn how infrared sensors work.

The three-wheeled robot, identified as Model 4Z6-911 LINE TRACER II, is offered in kit form with easy-to-follow mechanical assembly instructions. The kit offers a great deal of personal satisfaction and a sense of accomplishment after you have completed the approximately 2 hours worth of mechanical assembly required. All electronic

The LINE TRACER II, a low-cost robot from Stock Model Parts, uses an infrared sensor to automatically follow a black line guide path (courtesy Stock Model Parts).

elements are contained in two presoldered and pretested printed circuit boards.

The 5 1/2-inch-diameter, 2 1/2-inch-high robot consists of 55 major components plus over 130 fasteners. Among the major parts are two dc motors; rugged, yellow-tinted, molded body parts; plus a geartrain assembly. Also included is an open-end wrench and a tube of lubricant. The only additional tools needed are a small Phillips-head screwdriver and pliers.

The LINE TRACER II, part of the SMP/MOVIT series, can operate in a turning radius of as small as 6 inches in diameter. The robot's power source comes from one 9-volt and two AA batteries (not included). The LINE TRACER II robot kit is available at just $44.95 postpaid.

MEMOCON CRAWLER

The MEMOCON CRAWLER is more sophisticated than its scoot partner, LINE TRACER II, and costs a bit more: $79.95 postpaid. This low-cost, programmable, 4K RAM robot is controlled through its seven-function teach pendant. The robot includes an on-board, CMOS static RAM 256-×-4 sequencer. Applications include school science projects, robotics courses, or personal enjoyment. Of special significance is the ability to program the robot through any of the popular microcomputers (with parallel interface) to gain experience about how real

NOMAD is a robot, operated from a Commodore or Radio Shack Color computer, that has artificial vision (courtesy Genesis Computer Corporation).

robots are controlled.

The three-wheeled robot, identified as Model 4Z6-918 MEMOCON CRAWLER, is offered in kit form with four pages of easy-to-follow mechanical assembly instructions. As with other kits, you'll experience personal satisfaction and a sense of accomplishment after having completed the two hours of mechanical assembly required. All electronic elements are contained in two presoldered and pretested printed circuit boards.

Buttons on the teach pendant can be used to program the robot to go forward, right, left, pause, sound a buzzer, light a LED lamp, or repeat a program continuously. The 5 1/2-inch-diameter, 2 1/2-inch-high robot consists of 51 major components plus over 140 fasteners. Among the major parts are two dc motors; rugged blue-tinted, molded body parts; plus two geartrain assemblies.

Also included is an open-end wrench and a tube of lubricant. The only additional tools needed are a small Phillips-head screwdriver and pliers. The MEMOCON CRAWLER, part of the SMP/MOVIT series, is powered by one 9-volt and two AA batteries (not included).

How to Buy These Robots

To order either of these two robots, or for additional information, write or phone Stock Model Parts.

FRANK HOGG LABORATORY AND GENESIS COMPUTER CORPORATION

NOMAD is a splendidly descriptive name for the robot offered by two companies, Frank Hogg

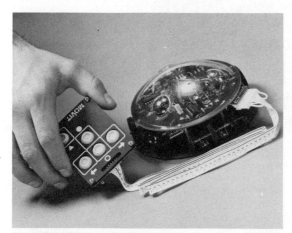
The MEMOCOM CRAWLER is a programmable, 4K RAM robot controlled through a teach pendant (courtesy Stock Model Parts).

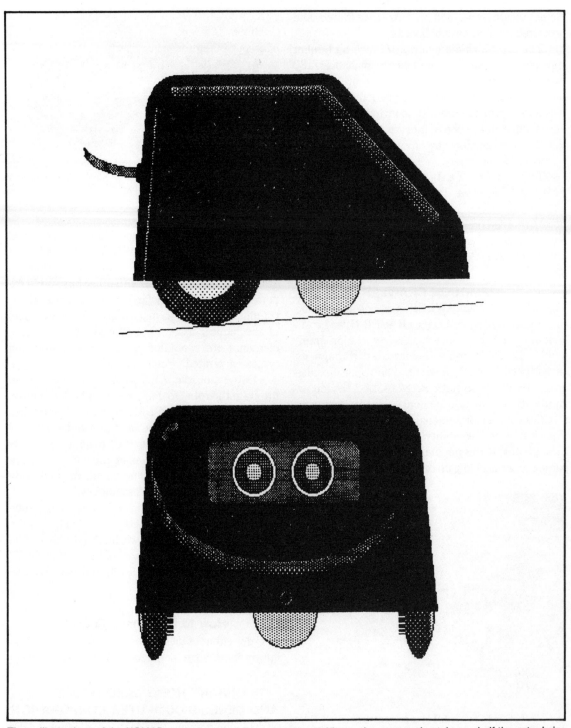

These illustrations of the NOMAD were generated by computer graphics and are approximately one-half the actual size (courtesy Frank Hogg Laboratory).

Laboratory (the TRS-80 color computer model) and Genesis Computer Corporation (for the Commodore 64 and VIC 20 computers). The rectangular-shaped NOMAD, 8 inches long, 7 inches wide, and 5 inches high, puts you on the leading edge with advanced experiments in robotics, artificial intelligence, and real-world applications. And does it move!

NOMAD can move forward and backward, and pivot left and right using two wheels driven by precision stepper motors. In addition, NOMAD can detect obstacles and map its surroundings with a specially engineered ultrasonic ranging system. NOMAD plugs into the computer's RS-232C port and has a 25-foot cord which allows for movement.

NOMAD's Body

NOMAD is manufactured on a sturdy aluminum chassis. A sturdy 1/8-inch thick, vacuum-formed plastic shell is secured to the chassis with five screws. A shallow cargo tray is molded into the top of the shell, and a rectangular cutout with filleted corners is milled into the front of the shell for the "eyes."

Inside the body are two 24-volt unipolar (8-wire) stepper motors which drive the two 2 3/4-inch wheels via precision injection, molded plastic gears. The "eyes" are ultrasonic transducers mounted in a foam-runner collimator, which is bonded to the inside of the shell. The motors and transducers are connected to a printed circuit board mounted on the chassis.

The PC board is connected via a 25-foot, 6-conductor cable (of the type used to connect office telephone sets to the wall junction) to an interface board housed in a standard ROM-pack cartridge. The interface board contains a PIA chip and some addressing circuitry. It is connected to a 24 Vdc, 400 ma transformer module which plugs into a 110-volt outlet.

NOMAD's Soul

NOMAD comes with a cassette tape which contains a machine language program called NOMAD. This program, which for the Radio Shack color computer will work with Color BASIC, Extended or Disk BA-

SIC, adds five more keywords to BASIC's set of commands. Four of these keywords are commands for controlling the robot; the fifth is a function for measuring the distance to obstacles.

NOMAD. The most important added keyword is NOMAD. This keyword allows you to move the robot in straight lines or circles, or to pivot in place. For example, to move forward 10 inches: **NOMAD 10,0.** To pivot right 90 degrees: **NOMAD 0,90.** To pivot left and move back to where we started: **NOMAD 0,90: NOMAD −10,0** To go around in a circle 36 inches in diameter, to the right: **NOMAD 36*3.1416,360.**

These examples demonstrate that the NOMAD command (like the other keywords provided) works exactly like any other BASIC command. You can use floating-point expressions for the arguments, put it on a continuation line, or whatever.

RANGE. The RANGE function is similar. This function, which requires no arguments (it works like MEM), returns the distance in inches to whatever obstacle it sees. An example of its use is the following one-line program which, despite its simplicity, provides some very interesting effects:

10 NOMAD RANGE-10,0:GOTO 10

This program measures the distance to an object (let's say it's your hand) and moves NOMAD to make the distance 10 inches. If he's farther than 10 inches away, he moves closer; if closer, he moves away. The effect is somewhat like an excited, curious, and somewhat wary puppy dog: It wants to get close enough to smell your hand, but not too close. I say excited because the RANGE function does not give the exact same reading every time, so that even if you hold your hand still, NOMAD hops back and forth by about an inch.

SPEED and ACCEL. You can modify the way in which the Nomad command works with the SPEED and ACCEL commands. The SPEED command sets the maximum speed of the robot in inches per second, and the ACCEL command sets the rate at which NOMAD accelerates to and decelerates from the maximum speed. A typical, general-purpose speed is 8 inches per second, and

Table 2-8. NOMAD Specifications.

NOMAD Specifications

Length: 8 inches	Weight: approximately 28 ounces
Height: 5 inches	Cord Length: 25 feet
Width: 7 inches	Cargo Tray: 4.25 × 3 × .25 inches
Wheel resolution:	approximately .04 in.
Ranging System: Resolutions:	approximately .2 in.
	Ultrasonic sight and motion detection capabilities
Maximum Speed:	About 17 in./second, depending on surface, load and slope
Maximum Load:	In excess of two cans of Coke
Maximum Slope:	About 40 percent (Coke will spill)
Interface:	Custom PIA board in ROM-pack case. Addressed at $FF70 (but modifiable). May be used with Multi-Pak, etc.
Power:	ac line

at this speed you can use accelerations up to practically infinity (such as 10000). At these very high accelerations, NOMAD tends to pop a slight wheelie when starting up.

If you decrease the acceleration (for example, ACCEL1), you can set the speed somewhat higher (SPEED 18 on a hard, level surface), and under these conditions NOMAD looks like he's pulling a train behind him. The effect is kind of weird.

NCONV. The remaining keyword, NCONV, allows you to change some of the floating-point conversion factors in the NOMAD software. You can set it up to use metric units (centimeters and radians) if you desire. You can also "calibrate" the ultrasonic ranging system.

NOMAD's Place in the Home

NOMAD is a lot of fun to use and experiment with. NOMAD comes with several example programs on tape; one of these, called "REMOTE," gives NOMAD a remote-explorer capability. You "drive" NOMAD using the arrow keys on the keyboard, and watch the screen where echoes from the ranging system are displayed. You can do a "radar scan" by pivoting NOMAD back and forth, and then steer in the direction where you don't see any obstacles. You can actually pilot the robot while it is completely out of your sight. The best part is, the source of the program is provided so you can modify it to suit yourself.

Buying a NOMAD

NOMAD for the Radio Shack color computer costs $250. That price includes software and a top-notch documentation manual. To order this model, or for more information, write or phone Frank Hogg Laboratory.

The NOMAD for the Commodore 64 or VIC 20 sells for $180. An optional BASIC enhancement language package retails for $39.95. To order or for more information, contact Genesis Computer Corporation.

Both Frank Hogg Laboratory and Genesis Computer are well-established, reputable firms which sell a wide variety of computer products and supplies in addition to NOMAD. When you write or phone, ask for a current catalog.

HOBBY ROBOT COMPANY

Hobby Robot Company of Hazlehurst, Georgia, offers a full line of robots, ranging from the SMART RABBIT, a kit model selling for as little as $329, to HECTAR, an industrial mobile robot that is priced at nearly $86,000! In between are such worthwhile robots as ART-II (cost: $10,950) and OSCAR (cost: $6,000). The founder and president of the company, Bill Dodd, is a man widely respected among robot researchers as a pioneer in the industry. In discussions with Bill, he told me that his robots are built as sturdy as possible and, with-

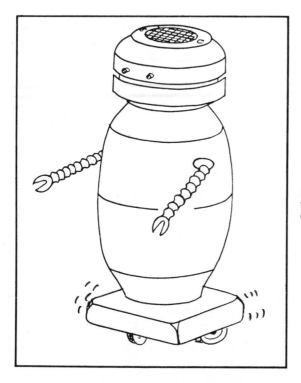

Hobby Robot's SMART RABBIT home robot interfaces with the most popular brands of home computers, including the Commodore models (courtesy Hobby Robot).

in each type and category, are the best to be found. Bill also says that Hobby Robot can build a custom robot to meet the user's needs.

Since Hobby Robot offers so many models, I'll focus here only on its SMART RABBIT model. This robot is not only inexpensive and affordable, but most definitely fits the definition of a home, or personal, robot.

SMART RABBIT

The SMART RABBIT robot is a personal robot that stands just under 22 inches tall and weighs less than 25 pounds. SMART RABBIT sports a strong, yet accessible, robot shell and powerful base gear motors. The SMART RABBIT is especially designed and priced as an entry level robot for young robotics enthusiasts. The SMART RABBIT will also be of major interest as a low-cost educational robot. The student is given the opportunity to learn electronics applications while building and testing the RABBIT. The hobbyist will find the SMART RABBIT of special interest because of its versatility and expandability.

The SMART RABBIT robot uses a novel control system allowing connection of almost any home computer to the robot. The SMART RABBIT can also carry an on-board Timex-Sinclair 1000 computer for independent control. The SMART RABBIT is available in the following two kit forms:

Level I Kit. The Level I SMART RABBIT is priced less than $350 and includes motorized base, head-mounted speaker and eye LEDs, and computer interface to allow control by either a Timex TMS-1000 (Sinclair ZX81) or a Commodore VIC or Pet 2000/4000 computer (via parallel output port). Interface to most home computers can be accomplished. Software is included. (Batteries and computer not included.)

Level II Kit. The Level II SMART RABBIT includes all Level I components and also includes motherboard, servo-powered arms and microgrippers, servo-powered head, servo-driver circuit cards, and the necessary mechanical components to provide a computer-controllable robot with functioning arms, grippers, and head. All servos are proportionally controlled via the computer bus (RDS44). The Level II kit comes with all components necessary to drive six servos and two dc motors. This complete kit is available for less than $800. (Batteries and computer are not included.)

Control System. The SMART RABBIT robot will interest computer buffs and model builders alike as it interfaces the computer to standard radio control (R/C) servos. The RABBIT's control system is a specially designed robot bus called the RDS44 (Robot Development System 44). The control system drives relays and servos via electronics mounted on standard (.125-inch space) 22/44 pin circuit boards. Actually, the RDS bus is expandable to a full 74 servo and 64 discrete relay output configuration. Discrete and analog input cards compatible with the RDS44 bus are presently under design and are planned to be available the last quarter of 1983.

Each SMART RABBIT kit is provided with a completely documented assembly and operations manual. The manual provides step-by-step assembly and testing procedures and provides sample programs and experiments that you will find useful in operating and upgrading your robot. A cassette tape is provided with sample software for running the robot (both Level I and II).

An additional feature of the RABBIT robot system is that it allows robot control via standard R/C transmitter/receivers. The RABBIT's servos may simply be plugged into an R/C receiver for proportional radio control.

The RABBIT is expandable to carry on-board screen and upgraded memory, and the RDS44 bus system will support discrete and analog inputs. Also under development, and compatible with the RDS44, are ultrasonic range-finder and infrared detectors.

How to Buy SMART RABBIT

To order the SMART RABBIT kits or for additional information about Hobby Robot products, write to Hobby Robot Company. Micro Management Systems, a company that sells computers and robots by mail order also offers SMART RABBIT. What's more, this company can sell you a SMART RAB-BIT model that comes fully assembled for $125 additional (for the level I model). For a catalog, write or phone Micro Management Systems, Inc.

CYBEROTICS

When Paul Mattaboni, formerly a staff engineer at M.I.T., decided to build a personal robot, he determined to do it right. Mattaboni, founder and president of Cyberotics, Inc., a fledgling Franklin, Massachusetts company, saw flaws in all the personal robots on the market and sought to construct a machine which would not have similar shortcomings. The results of Mattaboni's four years of research and development is CYBER I, a robot for security, industrial, educational, and hobby applications that uses an 8-bit 650Z processor with 30K of RAM.

CYBER I isn't your ordinary, run-of-the-mill dumb robot. He's richly blessed, with over 20 Polariod sonar sensors that provide vision. The company's talented engineer, Stephen Tougas, completed the hardware that accommodates this sophisticated artificial vision system. CYBER I can actually "see" obstacles and navigate them. He also has a form of artificial intelligence based on superior, newly patented software developed by Cyberotics.

Mattaboni expected to have the robot in full production by early 1985. Price should be about $3,000, including software for navigation. A gripper and arm will be an optional accessory.

An Introduction to CYBER I

Because Cyberotics has set out to push the technology of personal robots to its outer limits, it is worth our time to review the functions of this robot. The key features of CYBER I follow.

Ultrasonic Vision System. The ultrasonic vision system (patent pending) provides all the necessary data collection for a complete description of the robot environment. The vision system is used in both a collision avoidance role and as an environmental sensor system. The vision system relieves the user from having to program in the specific distances and directions for the robot to

CYBER I is a state-of-the-art robot that utilizes advanced techniques in artificial intelligence and machine vision (courtesy Cyberotics, Inc.).

traverse. The user need not provide the robot with a floor plan of its environment to prevent collisions with objects in a room. With CYBER I, collision avoidance is an automatic operation, thanks to the vision system.

Artificial Intelligence Software. An artificial intelligence software package (patent pending) provides the user with the ability to teach the robot in plain English using a library of behavior types. The CYBER I contains a library of over 100 possible actions or behavior/decision types, and the user has a catalog describing each of them. The user can describe a complex action or chain of actions in simple concise terms, providing the robot with a very user-friendly personality.

Cyberotics' philosophy has been to allow the software to carry the burden of sophistication. The software design of the CYBER I has applied the best elements of artificial intelligence programming without the need to use any custom or special languages, such as LISP. Instead, the company has concentrated efforts at using the public-domain language Forth as the high-level language and assembly language wherever speed is needed. Forth provides the necessary combination of compactness and ease of programming because of its object-oriented architecture.

The most impressive capability of the software design is the need for only an 8-bit computer to carry out all the actions of the robot in real-time. Cyperotics has designed the data structures for the artificial vision system and natural language processor to make maximum use of memory resources, while most of the robot control software is held in ROM without the robot.

Computer Compatibility. The CYBER I personal robot is powered by any of the major personal computers, from the Commodore 64 (for total autonomy) to the IBM PC (using a remote radio link). This is an important feature since the computer has other uses when the robot is not needed. It even allows the user to supply a presently owned computer.

Hardware and Physique. Because of the sophisticated software package, the hardware of the robot is simple and easy to understand or modify if desired. The hardware consists of a personal computer, an interface board, a motor drive control board for two dc motors, and a vision sensor control board whose output is simply a binary number denoting the distance to an object in a selected direction.

The hardware of CYBER I has been made as simple as possible, and, therefore, as reliable as possible. The motor drive consists of a VMOS power-transistor bridge-driver circuit which allows convenient motor direction control while using a single supply voltage. Speed control is provided by duty-cycling the motor control. The CYBER I does not use any kind of servo-feedback from the wheels because the artificial vision system provides all the necessary environment feedback. The vision hardware uses Polaroid Corporation ultrasonic sensors and their drive circuits, which are very complex, but inexpensive, units.

The final element of the hardware is the interface to the computer for all systems. The interface board connects the vision circuit board and the motor control circuits to the host computer's expansion bus. The board consists of connectors, buffers, and latches, as well as the motor control and speed circuits.

The physical robot is constructed with a strong, but light, sheet-metal base, which supports all the circuit boards and host computer and which has also been designed to eventually mount an arm with a fully extended carrying capacity of 10 to 20 pounds. A plastic shell is placed over the robot, the vision sensor assembly is mounted in the shell.

A Multipurpose Robot

As a mobile robot of top-notch technological capability, CYBER I is expected to have a number of applications in education, industry, and home. For example, some realistic applications would be:

☐ To transport parts and objects with the robot travelling between known points in a nonstructured environment, such as bringing parts to workers or to fixed robots on the factory floor.

□ As an industrial or home security device which requires varied behavior in a nonstructured environment, CYBER I can operate without human attention and prevent false alarms by having the ability to make intelligent decisions based on varied forms of sensed data.

□ In industry, for flexible materials storage and retrieval using existing storage bins and standard material containers.

To obtain additional information about CYBER I, write or phone Cyberotics, Inc.

EZ MOWER

"What can your robot do?" most people will ask of you after you bring your first personal robot home and start putting it on display. Usually, the new robot owner must duly admit that the robot has limited usefulness. After all, today's robots can't take out the trash, make the bed, or mow the lawn, right? Wrong! ROBOMOWER offers the busy homeowner the ultimate: a remote-controlled robot lawnmower.

The EZ Mower Company, the maker of ROBOMOWER, may just have a real winner here. At Albuquerque's First International Personal Robot Congress in 1984, inventor-designer Reza "David" Falamak, president of EZ Mower, won the distinguished Golden Droid Award for the "Most Useful" robot at the Congress. There, Reza was kind enough to give me a demonstration of his ROBOMOWER.

The impressive machine actually works—and works with precision. It is extremely easy to operate with its handheld remote-control transmitter. Reza told me that the ROBOMOWER makes life a lot easier for home owners weary of the ritual of lawn maintenance and saves industrial customers a bundle in maintaining the grounds adjacent to buildings and facilities. "Even more important," the enthusiastic inventor emphasized, "is the fact that ROBOMOWER will save 100 lives and prevent 115,000 injuries now incurred annually as a result of conventional mower accidents."

Unfortunately, as of late 1984, the ROBOMOWER was not yet in full production. The first ROBOMOWERs are expected to retail for $600 to $700.

I'm positive the ROBOMOWER will be of interest to personal robot enthusiasts. Let's review this unique robot in greater detail.

About ROBOMOWER

Robmower is a unique, remote-controlled lawnmower that is capable of cutting and trimming as well as shaping your lawn. It allows you to overcome the inconveniences of a conventional lawnmower while providing maximum safety and comfort.

Easy operation of ROBOMOWER makes it possible to be controlled by anyone who is able to operate a remote-control television. Complete operation is performed remotely. The operator controls this machine by using a simple device (transmitter) that fits securely in the palm of his hand. This transmitter contains four operating buttons and is capable of starting and killing the engine, stopping and driving the lawnmower, controlling speed, and automatic turning, which allows you to move the machine at any angle you desire, including a complete 360-degree turn on the spot without any turning radius and with maximum traction.

Safety has been a major determining factor in the design of this machine. Hands-off operation from activation to task completion guarantees the safety of the operator. The safety of the operation area is secured by a protective cover with "stop-on-touch" feature. If a person were to come in contact with the lawnmower while it is in operation, the "stop-on-touch" feature automatically stops the wheels as well as the blades from rotating, preventing any injury.

ROBOMOWER performs very effectively to eliminate a shocking number of injuries (77,000 from conventional lawnmowers and 38,000 from riding mowers, and up to 100 deaths) which occur each year. It also performs more efficiently than ordinary lawnmowers based on a maneuverable and quick-action platform with a unique multiple driving system.

Simplistic and inexpensive devices suitably used in the construction of ROBOMOWER make

Yes, Virginia, there is a robot that mows lawns: the amazing ROBOMOWER. In the bottom photo, Reza Falamak, inventor of this unique robotic device, sets the remote-control transmitter. The top photo shows the programmable ROBOMOWER navigating itself past an obstacle as it smoothly mows the lawn (courtesy EZ Mower).

it surprisingly affordable to the consumer.

EZ Mower, which has its motto the phrase, "Practical Robotic Technology," lists these outstanding advantages of its ROBOMOWER:

- ☐ 100 percent maneuverability in all directions, a feature which cannot be improved upon.
- ☐ 100 percent traction, greatly reducing chances of the machine getting stuck regardless of the condition of the lawn, another feature which cannot be improved upon.
- ☐ Zero turning radius, still another ROBOMOWER feature which cannot be improved upon.
- ☐ No match by present technology within its price range.
- ☐ Hands-off operation, which provides maximum safety.
- ☐ Convenient to the elderly, disabled, pregnant women, and teenagers.
- ☐ Size, load, and task adaptability.
- ☐ A mulching blade, which cuts grass into fine pieces, thus eliminating the need for a grass catcher, and fertilizing the lawn.
- ☐ A revolutionary concept which is affordable by the average consumer.

For More Information

To find out more about EZ Mower, Inc., its ROBOMOWER, or to order a machine write to EZ Mower, Inc.

ROBOT REPAIR

They call Gene Oldfield the "Robot Wizard," and rightly so. He has built dozens of robots and is constantly coming up with modifications and improvements to his creations. His company, Robot Repair, currently offers to robot fans a robot that Gene feels is his best to date: the ROBOCYCLE. It is definitely one of a kind. Before we look at the ROBOCYCLE, however, let's delve a little bit into the accomplishments of its maker—also the president of Robot Repair—Gene Oldfield.

Gene has a long history of achievement. From 1966 through 1969, he worked at the Lawrence Radiation Laboratory in Berkeley, California with the Nobel prize-winning Alvarez Physics Group. For many years, he taught math and physics at major colleges and universities and has served as high-tech or robotics art curator for museums and science exhibitions. The "Robot Wizard" built his first robot in 1978, a computer-controlled device called *Entropy*, and along with Izzy Schwartz, created over 400 pieces of electronic jewelry (yes, electronic!). Gene also was an engineer for Kylex Corporation in Silicon Valley, acting as project manager for a computer-aided design system.

In 1983, Oldfield created ROBOCYCLE, a robot that has gained much publicity. For example, Sacramento's Channel 5 TV station, on its *Evening Magazine* program, pictured the inventor and his ROBOCYCLE running side by side down a jogging path. The TV program characterized Oldfield as the "Henry Ford of the robot world."

The ROBOCYCLE

Now for a discussion of the ROBOCYCLE. As an educational robot, it is a high-performance mobile robot with sonar and digitized wheels that uses the SYM computer. It is designed to teach microprocessor feedback and the development of motion software. It is also designed to please a crowd, even with elementary programming.

Everything on the robot is exposed and functional. Weighing only 20 pounds, ROBOCYCLE goes better than 10 MPH and spins like crazy. Oldfield states that kids call it "bad," engineers call it "neat," and that it never fails to draw a crowd. "ROBOCYCLE," he adds, "is fun, educational and made to explore motion with a computer." The ROBOCYCLE is fast for a robot—about 44 times faster than RB5X and 10 times faster than Heath's HERO 1.

ROBOCYCLE is basic. It has no fancy skin—everything is exposed and ready for working on. The computer is the rugged and durable SYM, a popular and well-documented engineering computer often used in schools to teach microprocessor interfacing. The SYM has been modified to have nonvolatile RAM (no tapes!), and it has a resident (ROM) assembler editor. Among its 50 to 70 ID lines

ROBOCYCLE is the fastest mobile robot built. He can trek along at 10 miles per hour, yet has an excellent turning radius (courtesy Robot Repair).

is an RS-232C and a TTY. One bolt holds on the batteries, and all the power wires are visible. ROBOCYCLE is built for further development by the owner. It's a very tough development base. Oldfield says it has crashed into a wall at full speed and survived. Its high-tech bicycle wheels are unique among mobile robots.

For More Information

To obtain more information about ROBOCYCLE, or to order one, phone or write Robot Repair.

RADIO SHACK

Tandy's Radio Shack, the nationwide chain of retail

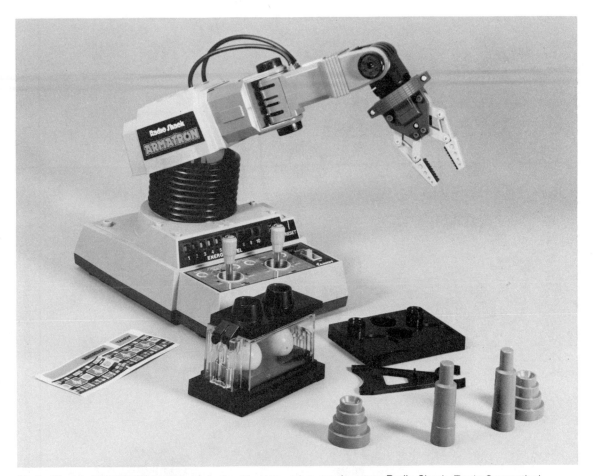

Radio Shack's ARMATRON is a fun and educational toy robot arm (courtesy Radio Shack, Tandy Corporation).

electronics stores, offers ARMATRON, a simple, noncomputer-controlled arm robot. This small arm is not designed for heavy-duty school use, but is acceptable for home educational use and as a home toy. Even with its limited usefulness, the robot isn't a bad bargain considering its affordable price of $20 to $30.

The ARMATRON comes with two control levers (joy sticks) and a game that allows you to move tiny globes and cannisters back and forth to a module (container). The arm has capability for movement and eight different positions.

ARMATRON can be purchased at any of the many Radio Shack stores. The robot comes with a complete 60-day warranty.

AMAROBOT

California's Amarobot, like Robotland in Florida, has big plans which involve the start-up of a chain of personal robot retail stores across the United States and Canada. Then, after North America? Well, there are 160 plus countries in the world, aren't there?

To fulfill his ambitious plan, Richard Amaroso, president and founder of Amarobot, has already taken two huge steps. First, he has opened up the first Amarobot Personal Robot Store in Berkeley, California. Second, he has already invented and is now manufacturing an entire line of patented personal robots. Among these robots are the first male/female pair of robots, MARTHA and

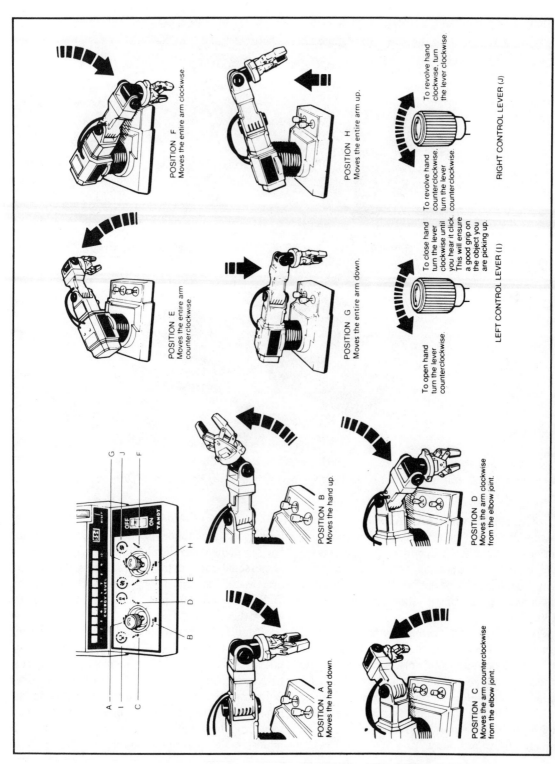

POSITION F
Moves the entire arm clockwise.

POSITION H
Moves the entire arm up.

To revolve hand clockwise, turn the lever clockwise.

To revolve hand counterclockwise, turn the lever counterclockwise.

RIGHT CONTROL LEVER (J)

POSITION E
Moves the entire arm counterclockwise.

POSITION G
Moves the entire arm down.

To close hand turn the lever clockwise until you hear it click. This will ensure a good grip on the object you are picking up.

To open hand turn the lever counterclockwise.

LEFT CONTROL LEVER (I)

POSITION B
Moves the hand up.

POSITION D
Moves the arm clockwise from the elbow joint.

POSITION A
Moves the hand down.

POSITION C
Moves the arm counterclockwise from the elbow joint.

The eight positions possible with the ARMATRON robot (courtesy Radio Shack, Tandy Corporation).

GEORGE, a one-of-a-kind HYDROROBOT, a TUR-TLE robot, and ISAAC, a full-size computerized robot available in kit format.

MARTHA and GEORGE

MARTHA and GEORGE are the pride of the Amarobot personal robot line. The company says that MARTHA and GEORGE have real personalities, complete with six male or female talking faces that appear on the 80-column, flatscreen display monitor. Here are the characteristics and specifications for this female and male duo, named, by the way, after two famous personages in American history:

☐ 80-column flatscreen monitor.

☐ Dual disk drives.
☐ On-board 600K RAM.
☐ 5 feet tall, 30 inches wide; built strong to work hard.
☐ Removable/interchangeable arms/end effectors; can be mounted simultaneously to perform a variety of functions.
☐ Tape deck/laser disk.
☐ AM/FM-TV.
☐ Wireless telephone.
☐ Real-time clock.
☐ On-board vacuum system.
☐ Cooks and delivers food.
☐ Home security system.
☐ Satellite communications link.
☐ Vision.

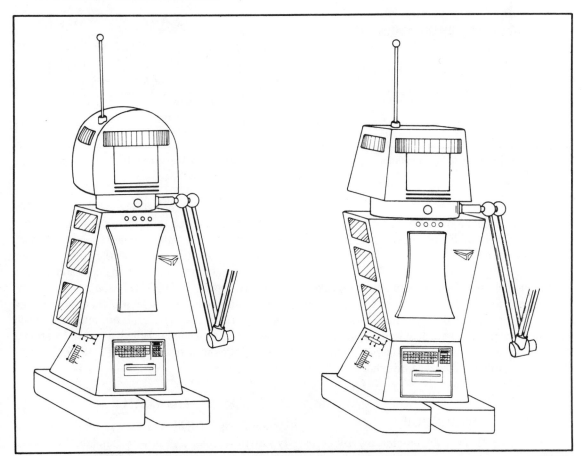

MARTHA and GEORGE, shown here without their end effectors, have multiple accessories, such as a tape deck, AM/FM radio, television, 600K on-board computer, clock, and home security system (courtesy Amarobot).

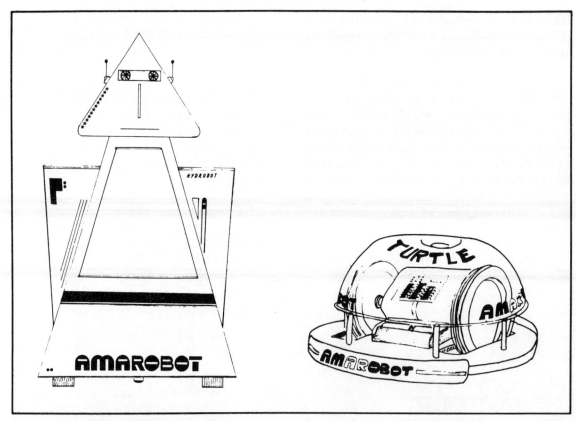

Amarobot's HYDROBOT (left) is a toy robot powered totally by water. To the right is Amarobot's TURTLE, an expandable, mobile robot (courtesy Amarobot).

☐ Voice.
☐ Voice recognition.
☐ Artificial intelligence/smart systems.
☐ Software.

As you can see, these robots are jam-packed with accessories and power. I haven't had the opportunity to see an actual demonstration, but if they meet up to the performance standards expressed by Amarobot, they are two fine humanoid machines. For the price, though, the standards should be high: they cost $8,500 retail.

HYDROBOT

The HYDROBOT is a dual-motor toy robot in kit form powered totally by water, using the miracle of shape alloy engines. The Nitinol alloy in the

engine expands and contracts power that allows this robot to run totally by the temperature variance between hot and cold water. No batteries are needed.

Simply pour hot water into the right shoulder and pour ice water into the left shoulder and instantly watch your HYDROBOT zoom off. Meanwhile, an on-board hydroelectric generator creates flashing LED displays, and an internal device creates engine sounds. Billed as an educational science kit as well as a fun toy, HYDROBOT costs $135.

The TURTLE

Armarobot's TURTLE is very removed from the typical turtle robot that is not expandable. This one, retailing for a basic price of $199, promises expandability. Most options are still under development. For now, an optional vacuum attachment ($35) is available. Planned are an arm, voice, vision, graphics, and many others.

ISAAC

ISAAC is a home robot kit with an arm that lifts 5 pounds, an on-board 16K computer, RS-232C option, and voice recognition option. The robot has a range-finding vision system and menu-driven software. ISAAC is said to be a hobbyist's delight with many expansion slots for development and user experimentation. Kit prices range from $799 to $1,899.

Buying a Robot From Amarobot

For more information (a catalog is $2 refundable), or to order Amarobot's robots, write or phone Amarobot.

TECHNICAL MICRO SYSTEMS

This company offers a turtle robot called ITSABOX, a highly appropriate name given the robot's boxlike shape. It literally resembles a rectangular box on wheels. The prototype of the robot has been featured in *Computers and Electronics, Popular Science,* and other magazines, and Lee Hart, president, tells me production is imminent.

The Intelligent ITSABOX

ITSABOX is designed especially for students, experimenters, and hobbyists. An intelligent turtle robot, it is a tabletop demonstrator of robotics principles and the capabilities of microcomputers for control applications. Although no bigger than a breadbox, ITSABOX has many of the capabilities of larger, more expensive robots and trainers. A computer on wheels, ITSABOX can move about, sense its environment, locate and manipulate objects, make decisions, and report on its progress. It is completely self-contained, and requires no connecting cables or remote computer. The on-board computer can be programmed in BASIC, Forth, or other high-level languages. An RS-232C interface is provided for such applications so ITSABOX can communicate with any standard computer or terminal.

The ITSABOX turtle robot was first conceived as a demonstration project to show the versatility of microcomputers in a control-oriented application. The original turtle robots by Terrapin Inc. pioneered the concept of the teaching robot, and offered a "hands-on" alternative to the conventional math-oriented approach to computer programming. The turtle was quite limited, however; it was essentially just a box with motors and "touch" switches, cable-connected to a remote computer. It had no vision, position sense, or pick-and-place capabilities.

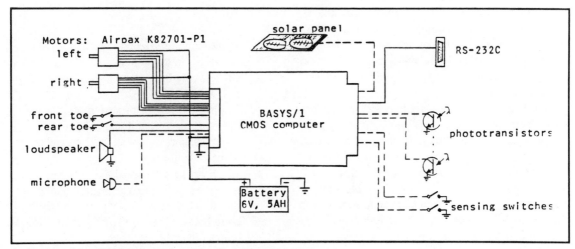

A diagram of the ITSABOX robot (courtesy Technical Micro Systems).

The ITSABOX is an intelligent turtle robot the size of a breadbox (courtesy Technical Micro Systems).

ITSABOX has taken the basic "turtle" concepts and expanded upon them. It is an intelligent, mobile computer with senses of touch and vision, able to make decisions and manipulate objects in its environment under program control. Its modular construction and simple design invites experimentation and demonstrates the concepts of control computers that are so often ignored.

ITSABOX lives in a two-dimensional tabletop world, especially convenient for instruction and demonstration. This "keep it simple" approach greatly simplifies the mechanical and programming details without sacrificing generality. ITSABOX can move about freely on its table, exploring as it goes with its vision and touch sensors for objects, walls, or table edges. Small objects can be picked up and moved with its hand, and rearranged according to your instructions.

Dissecting ITSABOX

Features.

☐ A self-contained robot, with its own on-board computer.
☐ High-level language programming in BASIC or Forth.
☐ Stepper motors for precise motion control.
☐ Pick-and-place manipulator hand.
☐ Vision sensors for object location and movement coordination.
☐ Touch sensors for hand and edge of table.
☐ Wireless (IR) link to teaching pendant or terminal (RS-232C).
☐ Long battery life: 1000 hours computing, 2 hours movement.
☐ Rugged, modular construction.

□ Generic 44-pin 4 1/2 × 6 1/2-inch cards for easy expansion.

□ Cost: $495 (kit), $595 (assembled). Parts may also be purchased individually.

Movement. Physically, ITSABOX measures 6 × 8 × 5 inches, and weighs under 4 pounds. It moves on two 3 3/4-inch tires, mounted on the left and right sides. Each wheel is driven by its own stepper motor for precise movement control. Stepping both motors ahead or back moves the robot the precise distance specified. Stepping them in opposite directions causes ITSABOX to turn exactly in place. On a smooth tabletop, ITSABOX can perform a complex series of moves and return to its starting point quite accurately.

Touch. ITSABOX has touch sensors for detecting table edge "cliffs" and barrier "walls." The "toe" switches are mounted under the front and rear edges of ITSABOX's chassis, and constantly check for the presence of the tabletop to the front and rear. Also, each finger of the hand has a "finger" switch to signal whether the hand touches an object or wall.

Vision. A series of eight phototransistors are mounted vertically between the fingers of ITSABOX's hand. Each one is focused so its range of vision intersects the tabletop a different distance from the robot. The closet is immediately in front, and the farthest about 12 inches ahead. ITSABOX can measure the light level of each in 256 shades of gray. By rotating left and right, a circular region can be scanned for objects. ITSABOX can thus locate dark objects on a light background, or vice versa, and move to position them between its fingers.

Hand. ITSABOX's hand has two horizontally opposed fingers. The fingers can be opened and closed by a stepper motor, and a pair of sensing switches are included to recognize when the fingers have closed on an object. The finger switches also double as wall sensors when moving and turning. Once ITSABOX has grasped an object, continuing to step the hand closed will lift the object off the table. ITSABOX can grip objects up to 2 inches in diameter and lift objects up to 4 ounces.

Sound. A sound generator can be pro-grammed to generate various tones and sounds through the built-in speaker. The speaker is also used with the teaching pendant to provide operator feedback on the robot's status. A speech synthesizer is under consideration for future expansion.

Communications. ITSABOX communicates with you via the teaching pendant or the RS-232C interface. Both methods are completely wireless and use an invisible infrared light similar to that used for television remote controls. The teaching pendant is a handheld keypad about the size of a pocket calculator. It can command the robot to move forward or back, turn, beep, and pick up or place an object. You can also program a sequence of commands that can then be executed as a single command.

The RS-232C interface is provided so ITSABOX can communicate with any standard terminal or computer, allowing you to develop much more sophisticated programs in high-level languages and run them in ITSABOX. Unlike the teaching pendant, the RS-232C interface handles two-way communications. Commands or programs can be sent to ITSABOX, and it can report back on their execution, error messages, or ITSABOX's view of its world.

Intelligence. ITSABOX's on-board brain is a BASYS/1 microcomputer, specifically designed for control applications. The standard BASYS/1 board includes 6K of memory and your choice of ROBOT BASIC or 8th (a variant of the Forth language) in ROM. Both include facilities for machine-level programming for extra flexibility. BASYS/1 is 100 percent CMOS logic for the ultimate in low-power operation and reliability. Power consumption is so low (under 25 mW) that ITSABOX's computer can remain active for weeks on a single charge, permitting even solar recharging.

Expansion. BASYS/1 also includes a breadboarding area for adding your own enhancements, and has an expansion bus for additional I/O and memory boards. Finally, ITSABOX comes with complete software listings and schematics, and user modifications and improvements are encouraged.

Specifications. Computer: 8-bit CMOS, using RCA COSMAC microprocessor, CDP1805, 2K user RAM, 4K ROM with Robot BASIC or Forth, 16 high-current outputs, 12 inputs, serial I/O port, 1 timer, 1 counter, 25mW typical power consumption.

Battery: 6V, 2AH rechargeable battery; ac adapter included.

Motors: 4-phase stepping motors—2 for drive wheels, 1 for hand.

Vision: 8-element in-line phototransistor array, with software A/D conversion to 256 gray levels.

Sensors: Two "toe" switches to sense table edge; two "finger" switches on left and right fingers.

Speaker: 2-inch loudspeaker, with software tone generator.

Pendant: 16-key handheld teaching pendant. Commands include move, back, left, right, open, close, beep, program, run, stop.

Serial: RS-232C standard, 300-baud, half-duplex IR link.

Bus: Standard edge connector, 44 pins on 0.156-inch centers. Expansion slot accepts standard 4 1/2 × 6 1/2-inch cards.

How to Buy ITSABOX

You can order ITSABOX direct from the manufacturer. The robot comes with a 90-day warranty. The company also sells many accessories for ITSABOX and various motherboards and expansion boards. You can also buy the ITSABOX part by part, which is an advantage for hobbyists building their own robot who wish to use only selected components of the ITSABOX.

Order ITSABOX or receive additional information by writing or phoning Technical Micro Systems, Inc.

ARCTEC SYSTEMS

Arctec Systems is a leader in the personal robotics field. Among its many products is Micro-Ear, a voice command system for robots and computers that permits you to train your machine to listen to and respond to a 256-word vocabulary chosen by you. Now, Arctec has brought to the market another innovative product, the Gemini home robot.

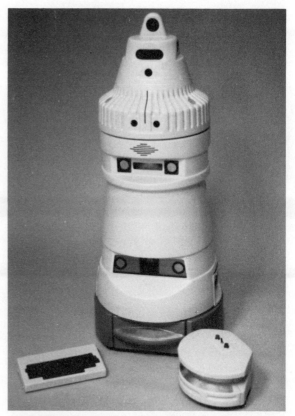

The Gemini robot system has 92K of software built in and two dc motors (courtesy Arctec Systems).

Raymond G. A. Cote, senior editor with *Robotics Age* magazine and an authority in the personal robotics area, says that Gemini is "the first of a new generation of robots." Says Cote, "Gemini is the first commercially available machine to actually move from place to place and go through doors without counting steps."

The Gemini can go through doors without counting steps and without driving your control software crazy or leaving it drained and ineffective. It does so by using infrared room beacons and passive infrared reflectors in the appropriate pathways of your home. The Gemini also does much more, however.

This is a powerful robot. It possesses two dc motors, a four-wheel drive, and three 12-volt batteries. Gemini will find its way around your home and office, perform preassigned tasks, charge its own batteries, talk to you, and obey your spoken commands. This life-sized robot is controlled by three on-board computers, an unprecedented 92k of built-in software, and a sophisticated array of electronic sensors all integrated together to give it life-imitating artificial intelligence. The moment you power up Gemini, the robot performs a self check of its hardware while verbally and visually informing you of the results.

Ready made, easy-to-use software allows you to demonstrate most functions of the robot with single keystrokes. Missions for the robot to accomplish at future times can be assigned by anyone. No programming experience is required. Important missions can also be accomplished, such as periodic reminders, wake-up calls, storytelling for the kids, remote control of appliances and more.

An enhanced floating-point BASIC language gives the user complete control of the robot. Commands like SPEAK, GOHOME, MOVETO, RANGE, etc., provide you with a means to develop complex programs, fast. Machine-language programs provide you with many debugging tools that you would expect to find on a real robot.

A high-level voice command language provides voice access for up to three users. The voice I/O and sound computer can store up to 256 words or phrases and dynamically update them as you converse with the robot. Highly accurate text-to-speech software and music programs with vocals are built in.

All in all, Gemini is a very advanced robot. Carl Helmers, publisher of *Robotics Age*, says it is "A significant step forward in the practical design of autonomous vehicles for research, practical personal use and experimentation." With endorsements like this, the Gemini should do well in the increasingly crowded market for personal robots.

The Gemini System

Gemini does a lot and so the cost is higher than that of many other robots. Following is a description and price listing for the basic robot, the robot enhanced as an educational system, and various accessories and software.

Gemini Basic Robot. The Gemini Basic Robot comes ready for navigation, obstacle avoidance, speech and sound generation, and voice recognition. Its artificial intelligence, named LMOS (for Life Mode Operating System), and scheduler are two of its advanced programs. It also includes sensors for temperature, motion, and light; a real-time clock; hardware random number generator; 8-bit A/D channels; a 40-character-×-8-line adjustable angle LCD; an unparalleled 3 on-board computers with CMOS Rockwell 65C02s for true multitasking; and a 56K RAM/ROM expansion card. The charger, two room beacons, and two door edge reflectors are also supplied. Both a User's Guide and a Technical Reference Manual complete the package. Price is $6,495.00.

Gemini Educational System. The Gemini Educational System includes the basic robot and the keyboard (can be used either infrared or direct connect), full floating-point BASIC, VOCOL, and the smoke detector. Price is $6,995.00.

Infrared/Direct Connect Keyboard. Store Gemini's full function keyboard conveniently on a shelf right below the LDD. Use with the infrared transmitter or telephone handset cord. It contains an HD6303 microcomputer with 2K ROM and 128 bytes of RAM. There are rechargeable batteries. Its price is $259.95.

Wafer Tape Drive. Use up to 128K of storage on this entrepo endless wafer tape. It comes with TOS, the tape operating system and is fast and convenient. Price is $179.95.

Infrared Motion Sensor. Located in Gemini's head, it detects heat and motion. It is good for security. Price is $119.95.

Room Beacons. Attractively packaged infrared room beacons expand Gemini's world. They use active infrared transmitters. You need one for each room. Price is $49.95.

Door Edge Reflector. Identify the doors for Gemini with these passive infrared reflectors. Two are needed per door. Price is $4.95.

Barometer. With a barometer and a little creative software, you can develop an expert weather system. Gemini can predict the weather for you by monitoring the trends in barometric pressure just like the weatherman. Price is $84.95.

Life Lites. Gemini's heartbeat, it is located in its chest, just like ours. It provides hours of enjoyment. Price is $49.95.

Smoke Sensor. It monitors visible and invisible particles in the air. It is UL approved. Price is $39.95.

Floating-Point Basic. Gemini's full floating-point scientific BASIC includes robot commands like SPEAK, MOVETO, HOME, and more. Price is $149.95.

Voice Command Language. It allows up to three users to command the robot. This may be used in either immediate or deferred mode. It features continuous adaptive learning by the robot. Price is $129.95.

Security Program. Feel secure when you're away from home with Gemini on guard. Preventive monitoring and/or action is taken if intruders are found. Price is $99.95.

8K CMOS Static RAM. Read and write memory expansion—6264LP12. Price is $39.95.

8K CMOS EPROM. Read only memory expansion—27C64. Price is $39.95.

Gemini Software Listings.

- ☐ Monitor RAM Listing: $39.95
- ☐ Navigation ROM Listing: 39.95
- ☐ Supervisor/Demo Listing: 39.95
- ☐ Scheduler Program Listing: 39.95
- ☐ Voice Command Language Listing: 39.95
- ☐ Security Program Listing: 39.95
- ☐ Procon Listing: 19.95

User's Guide. Easy to understand User's Guide helps you with Gemini step by step, from unpacking through using the scheduler, voice recognition, and BASIC software. Price is $29.95.

Technical Reference. Know Gemini inside and out. With this complete guide to its hardware and software, including circuit diagrams and software "hooks." Price is $49.95.

How to Buy Gemini

To order the Gemini robot or for additional information, write or phone Arctec Systems.

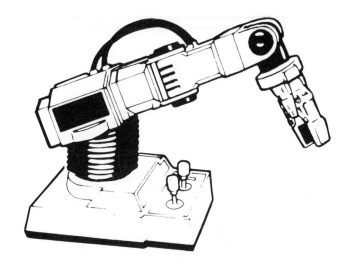

The Robotic Arm Educators

In the last section, we examined the many types and models of personal, or home, robots. Among them were a few robot arms, including the toylike AR-MATRON from Radio Shack and the highly functional robotic arms systems offered by Rhino Robots and Feedback, Inc. In this section, we'll look at several more robotic arm systems. All are sturdy, functional machines designed for educational and in some instances light industrial use.

If you're interested only in robots that physically resemble people and are mobile, these arm models may not excite your interest. With much of industry and factories across the world rapidly robotizing their operations, such systems are, however, the wave of the future. Because the ones featured in this book are designed primarily as educational tools, I believe it is worthwhile to briefly cover them. Addresses and phone numbers appear in Appendix B.

D & M COMPUTING

D & M Computing of Fargo, North Dakota is the distributor for highly acclaimed robotics training systems made by a British manufacturer. Among these systems are the ARMDROID I, retailing for $1,295 for the basic model, ARMDROID II ($4,000), and a mobile, turtlelike robot, the ZEAKER, selling for less than $200.

The ARMDROID I is a programmable robot arm that interfaces in an RS-232C port with most microcomputers; for example, the Timex/Sinclair 1000 and the Apple II (with the addition of a circuit board). The arm can be purchased inexpensively in kit form or fully assembled. It comes with a cassette program and two manuals. Accessories are many, including heavy-duty grippers, rotary carousel, and IBM software diskette.

For more information, write to D & M Computing.

LAB-VOLT SYSTEMS

Like D & M Computing, Lab-Volt Systems of Farmingdale, New Jersey offers several outstanding robotic training systems. Each is computer-controlled and comes complete with a comprehen-

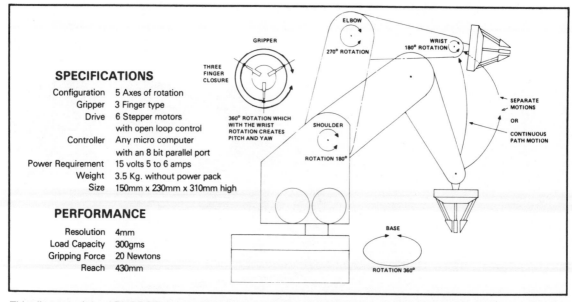

SPECIFICATIONS

Configuration	5 Axes of rotation
Gripper	3 Finger type
Drive	6 Stepper motors with open loop control
Controller	Any micro computer with an 8 bit parallel port
Power Requirement	15 volts 5 to 6 amps
Weight	3.5 Kg. without power pack
Size	150mm x 230mm x 310mm high

PERFORMANCE

Resolution	4mm
Load Capacity	300gms
Gripping Force	20 Newtons
Reach	430mm

GRIPPER

THREE FINGER CLOSURE

360° ROTATION WHICH WITH THE WRIST ROTATION CREATES PITCH AND YAW

ELBOW
270° ROTATION

WRIST
180° ROTATION

SEPARATE MOTIONS

OR

CONTINUOUS PATH MOTION

SHOULDER
ROTATION 180°

BASE
ROTATION 360°

This diagram of the ARMDROID I exemplifies the workings of a robot arm (courtesy D & M Computing).

ARMDROID I, the robot arm system from D & M Computing (courtesy D & M Computing).

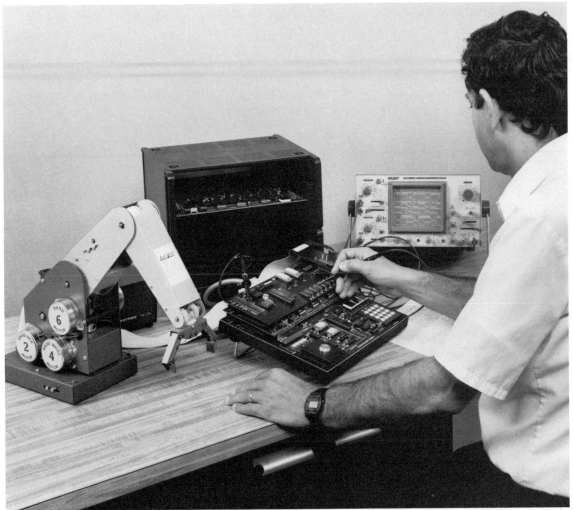

The Lab-Volt Robot Mechanism Training System (left) interfaced with Lab-Volt's Microprocessor Training System (center). The student performs the exercise outlined in Lab-Volt's correlated courseware using an oscilloscope and probe (courtesy Lab-Volt).

sive set of training materials. For someone interested in becoming a robotic technician, or for a school wishing to set up curricula, the Lab-Volt courses are ideal.

For more information, including a brochure outlining the Lab-Volt courses and equipment, write to Lab-Volt Systems.

PREP, INC.

Another robot educator is Prep, Inc., a New Jersey firm that distributes the SCORBOT-ER III Educational Robotics Program developed by an Israel company, Eshed Robotec. The Prep training system includes seven texts which comprehensively teach the fundamentals of robotics and also provide for advanced laboratory experiments. Student workbooks and cassettes supplement the instructional texts.

Prep's robot arm, the SCORBOT-ER III is a rugged-duty system that approximates larger industrial robots. Its RS-232 interface allows for com-

The SCORBOT-ER III education program offered by PREP, Inc. (courtesy PREP, Inc.).

munication with a host microcomputer.

For more information, write or telephone Prep, Inc.

MICROBOT, INC.

Microbot, Inc., a Silicon Valley firm, says that its TeachMover, priced at $2,595, is designed for engineers, educators, and hobbyists alike. TeachMover is self-contained, operated by a remote-controlled and handheld teaching pendant. Microbot's TeachMover robotic arm brings to robotics development its exclusive 13-mode "teach control." This simple set of buttons lets you program complex routines with ease, without having to learn languages or enter long commands.

The TeachMover has:

☐ Five axes of movement so that it closely simulates full-scale industrial robots.
☐ An on-board microprocessor with RS-232C

interface to allow operation with most computers.
☐ A built-in "intelligent" gripper that can sense objects and their size.

You can find out more about the Microbot system by writing or phoning Microbot, Inc.

REMCON ELECTRONICS

Remcon is now marketing a Teach Robot from Germany, where over 2500 of the robots have already been sold.

About the Teach Robot

Teach Robot is a precision-engineered, six-function electronic robot arm. The design incorporates many state-of-the-art techniques to give big robot versatility to a teaching robot. It was designed as a serious teaching aid for schools, colleges, and universities, as well as self instruction aid for individuals wishing

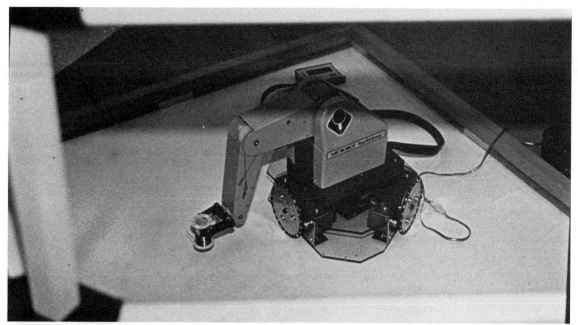

Microbot's TeachMover robotic arm comes with a "TeachControl" remote-control pendant (courtesy Marilyn Chartrand).

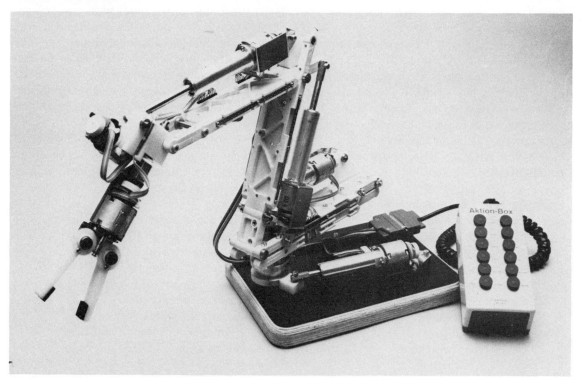

The Remcon Teach Robot is a German-built machine. Over 2500 of the robots have been sold in Europe (courtesy Remcon Electronics).

to learn more about robotics. Industry has also found use for Teach Robot in a variety of different roles not requiring a larger robot.

Mechanical Details. The robot was designed to combine metal with the properties of a new, light and extremely durable plastic. Teach Robot is constructed of injection-molded Delrin plastic with special metal brushings molded in at all joints and wear points. The components are assembled together with cap head screws. Future modifications of the robot (by the end user) have been considered, and prethreaded brass strips have been incorporated into the design along each molding.

The robot's six functions are controlled by six custom-made dc motors with shaft encoders. The optical shaft encoders allow the robot to be connected to an optional computer to precisely make repetitive movements. The motors drive the robot's movements through several different types of gearing. Wiring for the robot's motors come from the factory pretested and with prewired pin connectors for quick service removal. A spare pin is also provided at each harness for further customer modifications. A 25-pin Cannon connector is also provided to quickly connect the robot arm to its optional computer or to the optional action box controller.

The Teach Robot is shipped as a kit in a compartmented polystyrene box. It comes with a complete 90-page lay-flat instruction manual and takes approximately 6 hours to assemble.

Teach Robot Electromechanical Specifications:

- ☐ Number of drive motors: 6
- ☐ Voltage: 12-19V dc
- ☐ Medium load running current: 300 ma
- ☐ Shaft encoding: optical
- ☐ Rectalinear actuator speed: 120 mm/s
- ☐ Vertical reach maximum 760 mm
- ☐ Horizontal reach maximum 460 mm
- ☐ Angles of movement:
    Body: 90 degrees
    Shoulder: 90 degrees
    Wrist: 90 degrees
    Wrist roll: 540 degrees
    Claw opening: 180 degrees

- ☐ Positioning repeatability: + or − 0.5 mm
- ☐ Electrical termination: 25 pin Cannon connector.
- ☐ Weight (including base): 2.5 kg

Options

Action box. The action box connects to a dc power source of 12 to 20 volts (at 1.5 amps). The action box is a series of control switches necessary to supply current to the arm's electronic motors. The action box is wired to give dynamic breaking on release of each switch. The action box connects to the robot through the Cannon pin connector. The action box does not, however, take advantage of the shaft encoders wired into each of the motors and is only used for driving the robot arm.

Drive Interface. Now being designed and not yet available are six Euro-boards, double-sided with t.h.p. required to interface the Teach Robot with a computer. Each board drives one function of the Teach robot and carries circuitry to give two adjustable speeds of operation (fast and creep), as well as counter output of actuator position and state. The system bus (detailed in the manual) allows for successively programming the boards. The system bus can serve up to nine boards, giving many options for further use (stop switching, encoders, and indicator devices for training, etc.).

Drive interfaces are available in kit form: all components, printed circuit boards, and hardware are included, except for the integrated circuits, which are universally available.

- ☐ Number of boards: 6 Euro-Boards
- ☐ Operating voltage (Logic): 5V dc
- ☐ Operating current (Logic): 200 ma
- ☐ Operating voltage (Drives): 24v dc
- ☐ Operating Current (Drives): 300 ma
- ☐ Counter capacity: 64K
- ☐ Adjustable speeds: 2
- ☐ Indicators start/stop: 2 × LED red
- ☐ Signal level: TTL/LS

Computer. A ready-built and tested computer on a Euro card is available for the Teach Robot. The computer is based upon the 8085 processor, EPROM, and CMOS RAM ready to be mounted into one of the 11 places on the bus board. The computer contains a monitor program which enables the operator to program the Teach Robot with the Teach Box. Memory retention is by on-board battery.

- ☐ Operating voltage: 5V dc
- ☐ Operating current: 300 ma
- ☐ Output ports: 8 + 4
- ☐ Input ports including key port: 16

Teach Box. Teach Box comes ready-built and tested with a spiral cable and 25-pin Cannon connector to fit into the computer port. It is capable of programming all axes, memory recall/clear, etc. from its 29 function switches.

How to Buy the Teach Robot

The Teach Robot comes in kit form and sells for $377.50, exclusive of options. For the prices of options or for additional information, write or phone Remcon Electronics.

4

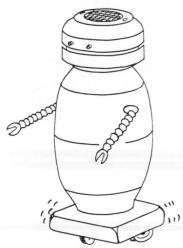

The Robots
Keep Coming:10 More Models

In this section, 10 more robots and robot kits will be briefly introduced and described. Photos are not available for all these models, and the information here is not comprehensive. So if what you read here catches your fancy, you may wish to investigate a little deeper by phoning or writing the company. In one case, Robotix™ robot kits, you can check out the products at your local department, toy, or hobby store. Addresses and telephone numbers appear in Appendix B.

MILTON BRADLEY

Milton Bradley is one of America's largest toy manufacturers. A past product was BIG TRAK, a small, mobile platform that many personal roboticists used for hobby projects. BIG TRAK was retired last year and replaced by two new Robotix series kits. Each is unique. Designed for kids ages 7 to 14 (or adults who enjoy these types of things, like this author), the kits are fascinating. For example, you can build a robot arm much like the Radio Shack Armatron, and you can build a variety of different sizes and shapes of rover walking vehicles. A gripper mechanism also comes with each kit, as do a two-legged walking assembly and a battery pack.

The Robotix kits are of Lego-type construction sets and are extremely educational. Assembly is quite simple and is not all that time-consuming. The kits are a great way to learn the principles of robotics and conduct artificial intelligence experiments.

You can connect either of these kits to your computer or they can be controlled from another external source. One company, Crabapple Systems, offers the KELP interface card which allows you to interface Milton Bradley's Robotix models with the Apple II, II+, and IIe computers. The card comes with software and documentation. (See "Supplies Section," Part III of this book, for Crabapple's address and phone number).

The two kits and their prices are as follows:

RM 1000: $39.95
RM 2000: $59.95

The higher priced kit has four motors vs. only two for the RM 1000, and this is a decided advantage for hobbyists.

ACORN COMPUTERS

The United States distributor of the British Acorn Computers Corporation is now offering the Cyber 310 arm robot, called the FORTH robot, which interfaces in English and has 5 degrees of freedom plus a gripper. The robot, which mimics larger industrial robots, is designed for robot education and hobbyist use.

Price is $1,399.00. The Cyber 310 is compatible with Acorn computers, Commodore's VIC 20, Apple, and other computers. For more information, call or write Acorn Computers Corporation.

ROBOTEX

This company offers a number of robotic items, including the RM-501, a small, portable robot arm programmable in BASIC via your personal computer. Robotex also sells RTROL, a robot programming language. For more information, call or write Robotex.

HOB-BOTS OF HUDSON

Hob-Bots of Hudson offers low-budget robot kits and assembled robot models, including MicroDroid and Sentroid. The robots resemble trash cans on wheels. The price for these two is $495.00 and $595.00.

CHESSEL MICROPRODUCTS

This English company markets a German-built robot plotter, the GR-15, that interfaces with computers with RS-232C or RS-423 standard serial interface. The desktop turtlelike robot, barely 4 inches square, is driven from a control unit and has a ribbon cable that allows full freedom of movement.

The robot plotter, only 3 pounds, scoots around on a sheet of paper and uses one of its three colored felt pens to make high-resolution images. It can make letter characters and draw circles, pie graphics, and other drawings and shapes.

Because this is an English company, you could have problems in getting the unit serviced and repaired. Still, it's an interesting device. Cost is about $375.00. To buy the GR-15, write Chessel Microproducts.

AMERICAN MICRO PRODUCTS

According to reports, this firm also has plans to market the German-built robot plotter just described. For more information, contact the company.

ECONOMATICS, LTD.

This company offers another British robotic device, the BBC Buggy, a tabletop, rectangular-shaped robot driven by two precision stepper motors. The small (4-inch-square) robot has detectors to feel objects and seek out light sources. It can be fitted with a pen drawing mechanism and a robot arm and is compatible with the Acorn and other personal computers. Ask for the current price.

HARVARD ASSOCIATES

This American company imports robot products from Australia and elsewhere. Among the robots for sale are the Robotnick P/L robot arm, the Tassman Turtle, and the Turtle Tot. The Turtle Tot is a small, programmable robot that moves, draws with a pen, and has various sensors. It interfaces with Apple, Atari, TRS-80, and IBM computers and retails for $325.00 to $400.00. Items are either assembled or in kit format. For more information, contact Harvard Associates.

EASTERN MACHINERY AND MANUFACTURING

This company's robotics division has developed a unique robot system, the Q-Bot kit. This kit allows the hobbyist to creatively build an arm robot with gripper or a variety of other devices. What makes it unique is that it comes unassembled, resembling a mechanical/electrical tinkertoy outfit. Parts include a motor, four wheels, light sensor, bases, angles, braces, struts, screws, battery pack, magnetic sensors, and bear boxes. With the Q-Bot kit, the hobbyist can build an automated cart, a unique arm robot, or an automated vacuum cleaner. Q-Bots operate on any computer with an RS-232C port. For more information contact the Robotics Division of Eastern Machinery & Manufacturing Company.

AXLON

Axlon is a new company founded in August, 1984 by Nolan Bushnell. Bushnell, you will recall, is also the founder of Androbot and, before that, Atari. He is also known as the inventor of the first video game, PONG.

Axlon is producing a robot that is a look-alike of F.R.E.D., of Androbot. Called ANDY, the robot is a compact, desktop model, 9 inches wide at the front view and 13 1/2 inches in stature. ANDY comes with PERSONALITY EDITOR and a sample BASIC program on disk. You can control him with the PERSONALITY EDITOR or from BASIC, LOGO, or Forth. The robot retails for $119.00.

Andy has a built-in sound generator and light, sound, and bump sensors. Reportedly, the software accompanying ANDY gives him different moods and tasks. He uses a "D" cell battery that lasts about 7 hours. This robot is for the Atari 800 or Commodore 64 computer.

Axlon also plans to introduce other robots. I have not seen ANDY demonstrated nor seen these other models. They may or may not perform on a par with Androbot's robots. Given Bushnell's past few years of experience with Androbot, however, these robots should be improved and enhanced in comparison.

For more information, write or phone the San Jose area company.

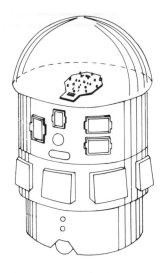

5

Custom-Built Robots

Okay, so you've looked over the gallery of personal robots presented here and you just can't find the one robot especially for you. Maybe you have your own unique idea about what your own robot should look like, or perhaps you own or manage a retail business which would benefit from having a promotional robot tout and publicize its products or services. Indeed, it might be that the needs of your business or yourself could best be served by a robot built to look like a TV set, bowling pin, or—why not?—a shapely blonde model!

Several robot manufacturers specialize in meeting individual needs by building robots to customer specifications. These companies also lease or rent robots to customers, complete with human operator. We will briefly discuss the work of four such manufacturers, each well-known and highly regarded in the industry. These are companies that have built custom-made robots for companies like Time-Life Books, Dayton and Macy's department stores, Heinz (the tomato ketchup people), and movie and television producers. Promotional robots are on display around the world, and their antics have been enjoyed by millions.

A word of caution, however: custom-built robots naturally cost more than mass-produced models. The typical price of individually designed models ranges from, say, $6,000 to $75,000 and up, depending on the materials that go into the robot and the labor involved. Also, most custom-built robots do not come with computers on-board. In fact, most are not even usable with computers; that is, they cannot be interfaced with computers nor computer-controlled. (An exception is the DC-2, from Android Amusement.) Instead they are controlled by hand-held radio transmitters. Maybe the robot purist might contend such creations are not "true" robots. Don't tell that to the droves of enthusiastic fans of these "showboat" electromechanical people. To them, a robot is a robot, and any robot is a fantastic marvel of modern high technology, embodying the very essence of man's future world.

Addresses and telephone numbers for these companies are provided in Appendix B.

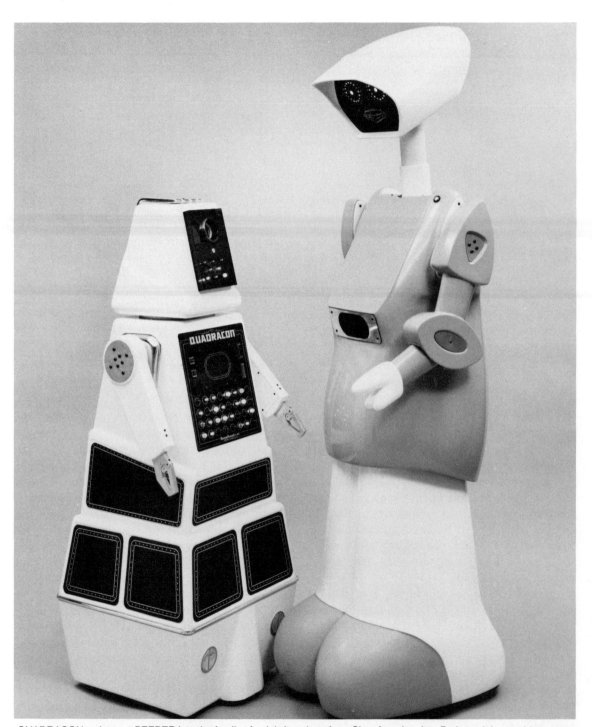

QUADRACON welcomes PEEPER into the family of celebrity robots from ShowAmerica, Inc. Both models are able to move wirelessly around an exhibit area or across a stage. They are able to shake hands and converse freely with astonished spectators (courtesy ShowAmerica Inc.).

Expo Ernie is one of the more unique robots developed by ShowAmerica. Created for the Canadian government and premiered at the recent World's Fair in Knoxville, Tennessee, Ernie is slated to be the mascot of the World's Fair of 1986 in Vancouver, B.C. The robot is presently on a national tour across Canada to build awareness for the fair (courtesy ShowAmerica, Inc.).

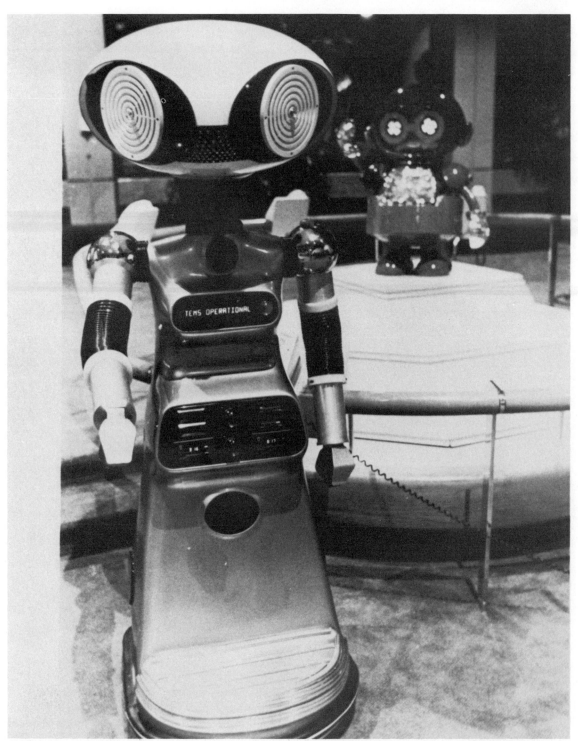

SICO (left) and SMART-1 (right, on pedestal) at Walt Disney's Epcot Center (courtesy Internatonal Robotics Inc.).

International Robotics can build a robot to your specifications. This is an artist's conception of a mobile beverage dispensing robot (courtesy International Robotics Inc.).

One of the popular robots from The Robot Factory shakes hands with an actor on a television show (courtesy The Robot Factory).

138

SHOWAMERICA INC.

John Anderson, vice president of ShowAmerica, says his is the "world's oldest and largest promotional robot company." The mature firm certainly has an impressive backlist of clients, including the Knoxville World's Fair, Exxon, Kodak, DuPont, Heinz, General Electric, Nike, Hoover, BMW, and . . . well, the register of happy customers is practically endless.

ShowAmerica clients have a choice of two standard models, "Quadracon" and "Peeper," in addition to the option of having a unique robot or product replica robot built to their specifications.

INTERNATIONAL ROBOTICS

International Robotics may not be as large as ShowAmerica, but the company has developed an exceptional reputation for its robots. One robot, SICO, has literally been all over the world, riding on airliners as a "human" passenger, entertaining flight attendants and fellow travelers alike. SICO has promoted many a company's products (Pfizer, Sperry Rand, and others) and for a while was a star on a TV soap opera where he became known for his patented sense of humor. SICO has been photographed with dozens of celebrities, from Prince Albert of Belgium to baseball great Joe DiMaggio and former president Gerald Ford.

International Robotics will build a robot to meet any customer's needs. The company will also construct entire theme exhibits featuring robots. For example, a recent client was a night club for which International Robotics provided an actual working model of a U.F.O. plus several futuristic robot waiters and waitresses.

THE ROBOT FACTORY

The Robot Factory of Cascade, Colorado, has for many years been a provider of robots for exhibits, company promotions, movie and TV productions, and individuals. Its first creation, an 8-foot-tall robot named COMMANDER ROBOT, made its debut as an ice skater for the Follies in 1966.

In addition to walking, talking electromechanical robots of plastic and metal "skin," the Robot

Bicycle riding robots with fur! The Robot Factory's HOT TOTS can perform with or without a human operator (courtesy The Robot Factory).

Factory also offers the animated HOT TOTS, fantasy fur creatures which can wear a variety of costumes and which thrill and delight youngsters (and adults!). They even ride bicycles. You have to see them to believe it. These robots are human-controlled and also come preprogrammed on 60-minute cassettes; so they can do their thing without a human operator.

ANDROID AMUSEMENT

Gene Beley, owner and co-founder of Android Amusement Corporation, Irwindale, California, is a pioneer in the world of robotics. In addition to his company's promotional robots, Gene is currently working with Ralph McQuarrie, the creator of

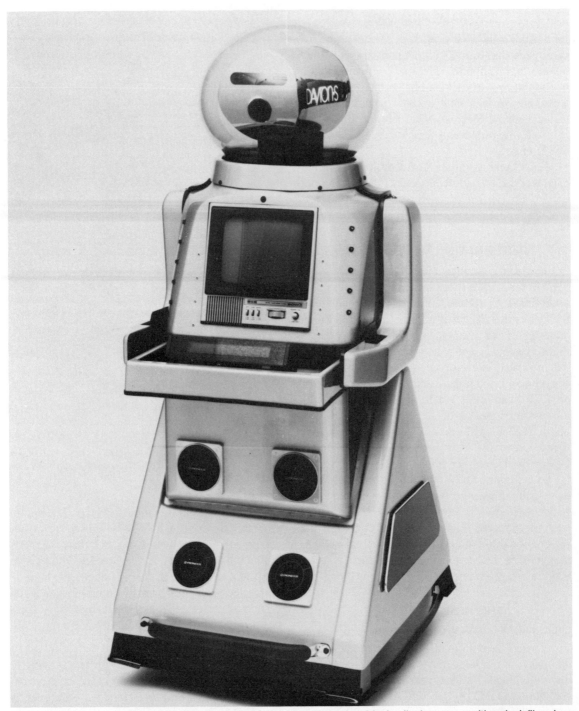

DC-2 makes frequent appearances on television shows like *Chips*. The 4-foot 2-inch tall robot comes with a sleek fiberglass body; green, flashing electronic displays; a serving tray, and a video screen. He can talk, sing, dance, and serve refreshments to guests (courtesy Android Amusement Corporation).

R2D2, to create a mass-market personal robot. Maybe I can include information about the result of their enterprise in the next edition of this book. For now, however, let's discuss the able promo robots that Gene Beley and his firm market.

Android Amusement's DC-2 robot has gone all over the world. This robot, which can be ordered in a configuration compatible with Commodore computers, has performed at a luxury hotel in Jakarta, Indonesia; another stint was as a mobile slot machine at a casino in Las Vegas. (The robot always shook hands with the player, win or lose!) DC-2 has appeared on such TV shows as *CHIPS* and *The Richard Hogue Show.*

The company also leases and sells two humanoid robots, ANDREA and ADAM, whose physical features truly personify feminine beauty and masculine good looks.

Part III
The Hobbyist Corner: Parts and Software for Robot Builders and Enthusiasts

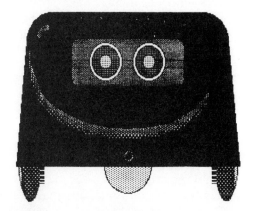

6

Attention Robot Builders and Enthusiasts!

In this section, we will review the products of a number of suppliers that offer robot parts and/or software. Most software available today is designed specifically for the HERO I, the most popular and well-known personal robot. As the field of personal robotics grows, however, so will the number of entrepreneurial software companies marketing products.

If you are seriously considering building your very own robot, the suppliers listed here are ideal sources. Stock Drive Products, for example, has over 24,000 small mechanical parts listed in its master catalog. The company provides drive components to nearly every major robot manufacturer. Other companies, such as Omega Engineering, also offer a wide variety of parts, and others, like Gloucester Computer, specialize in major accessories and components.

TO BUILD A ROBOT

Building a workable robot, especially a computer-controlled machine, isn't an easy task. As Raymond Cote, editor of *Robotics Age* magazine, recently noted, "The knowledge required to produce functional robots encompasses several different fields of knowledge." Chief among these fields is electronics and sensor operation, computer programming, and mechanical principles.

Still, with diligence and effort, a turtlelike robot can be constructed for as little as $25, a larger robot for no more than a few hundred dollars. Imagine the shock and amusement on the face of friends and family when they see your electromechanical person barging around the room or down a sidewalk, cracking jokes and gleefully greeting on-lookers! With every challenge comes rewards, and the rewards to the robotic experimenter can be great.

It is a difficult task to build a robot, then interface it with your home computer. Remember though, that hundreds, perhaps thousands, of people are doing it today, and have confidence that you, too, can construct a simple robot. Recently, Richard Prather, founder of the Homebrew Robotics Club in Palo Alto, California, addressed a Robotics Society of America group on the topic of building a robot. Prather, himself a robot builder, said that besides the technical aspects of putting a working robot together, the average robot hobbyist or experimenter must work within constraints of time and money. The number of hours you can devote and the number of dollars you can invest should be budgeted in regular, planned amounts.

For example, Prather said that it is more efficient—and wise—to put in 1 or 2 hours every day than a 10-hour stint every 2 or 3 weeks. The latter plan would be very fatiguing and also would not allow time for your mind to properly evaluate each step and its results. Also, with the longer time period, if you are short of cash, you could be saving money as you go along to buy parts and supplies.

To build a computerized robot requires a variety of different skills, including the computing for the control software, knowledge of basic electronics and motors, and the ability to mechanically construct the base and the body. Building a radio-controlled robot that is not computerized is a little more simple.

All the parts necessary to build a robot can be found at local hardware, electronics, and computer stores. The hardware store will have casters, wheels, wood (or metal), screws or bolts, and the other items you need to construct the robot's anatomy. At the electronics shop, you can buy inexpensive, small, low-voltage dc motors, for example, a motor that is run on 5 or 6 volts at 750 milliamps and can be conveniently powered by a rechargeable battery. For a large robot, some experts recommend wheelchair-type motors which are powerful.

It is wise to attempt a simple robot the first time. This robot could do a few useful tasks and demonstrate its limited talents to your family and friends—such talents as serving drinks, and entertaining guests with a few dance "steps!"

Design your robot carefully. The more planning you do, the better and more capable the robot will be.

SOME ROBOTICS BASICS FOR BEGINNERS

It isn't the purpose of this book to present schematics, plans, and detailed instructions for building robots. Nor do I have the space here to extensively discuss the principles of robotics. It may, however, be worthwhile to hit on just the basics of robotics for readers who are just beginning their study of this fascinating field. Presented here, courtesy of the National Aeronautics and Space Administration, is a "robot primer," which includes a discussion of all classes of robots.

A typical robot consists of one or more manipulators (arms), end effectors (hands), controller, power supply, and possibly an array of sensors to provide environmental feedback. Because the majority of robots in use today are for industrial purposes, classification of them is based upon their industrial function.

Robot Classes

Non-servo robot (pick-and-place robot)—the simplest form of robot. This robot picks up an object and places it in another location. Freedom of movement is usually limited to two or three directions.

Servo robot—robots of several categories that employ servomechanisms for the manipulator and end effector to alter direction in midair without tripping a mechanical switch. Five to seven directions of motion are common, depending on the number of joints in the manipulator.

Programmable robot—A servo robot that is driven by a programmable controller which memorizes a sequence of movements and repeats them perpetually. This kind of robot is programmed by "walking" the manipulator and end effector through the desired movement.

Computerized robot—A servo robot run by a computer. This kind of robot is programmed by instructions fed into the controller electronically. "Smart" robots, as they are known, may include the ability to improve upon their work instructions.

Sensory robot—A computerized robot with one or more artificial senses to sense its environment and feed back information to the controller. Senses are usually sight or touch.

Assembly robot—A computerized robot,

probably with sensors, that is designed for assembly line jobs.

Primitive Robotic Functions

Robots are primarily designed for manipulation purposes. The actions produced by the end effector or hand on the end of a manipulator are:

- ☐ Moving:
  - — from point to point
  - — following a desired trajectory
  - — following a contoured surface
- ☐ Changing orientation
- ☐ Rotation

Non-servo robots are capable of point-to-point motions. For each desired motion, the manipulator moves at full speed until the limits of travel are reached. Non-servo robots are often referred to as "limit sequence," "bang-bang," or "pick-and-place." When non-servo robots reach the end of a particular motion, a mechanical stop or limit switch is tripped, stopping the motion.

Servo robots are also capable of point-to-point motions, but movements of manipulators are accomplished with controlled variable velocities and trajectories. Motions of servo robots are controlled without the use of stops or switches.

Manipulator Arm Configurations

To accomplish robotic motions, four basic configurations of manipulator arms have been devised. They include rectangular, cylindrical, spherical, and anthropomorphic (articulated or jointed arm). Each arm design features two or more *degrees of freedom* (DOF), which refers to each direction an arm is capable of moving. A simple linear or straight line movement is one DOF. To follow a two-dimensional curved path requires two DOF, both up and down and right and left. More complicated motions require many degrees of freedom. To locate an end effector at any point and orient the effector in a particular work volume requires six DOF. More degrees might be required if the manipulator has to avoid other equipment or obstacles in performing its tasks. For each DOF one joint, rotary or linear, is required. to increase the versatility of the manipulator arm design, two or more of the four basic configurations can be combined on the same manipulator.

Actuators (Power Supply)

Moving the manipulator joints is accomplished by actuators. The three main forms of actuators in use today include pneumatic, hydraulic, and electrical. *Pneumatic actuators* are pressurized gas to move the joint. Gas is propelled by a pump through a tube to a particular joint and actuates movement. Air brakes on diesel trucks are a common form of pneumatic actuator. Pneumatic actuators are inexpensive and simple, but their movement is "squooshy," and consequently they are usually reserved for "pick-and-place" robots.

Hydraulic actuators are the most common industrial system and are capable of producing high power. The principle disadvantages of hydraulic actuators are the accompanying paraphernalia, including pumps and storage tanks, and problems with system leaks.

Electric actuators offer clean movements, can be precisely controlled, and are very reliable and accurate, but do not deliver the high power-to-system weight ratios that hydraulic actuators can deliver. For some actuator functions, electrical actuators are preferred because the cost advantage of hydraulic actuators diminishes with decreasing size.

Mobility

Most industrial robots are fixed in place or ride along rails or guideways. Other robots are built into wheeled carts. Still other robots use their end effectors to grasp handholds and pull themselves along. Very advanced robot mobility systems use articulated manipulators as legs for walking motion.

End Effectors

The "hand" or gripping device is usually attached to the end of the robot's manipulator. The functions of the end effector include grasping, pushing and pulling, twisting, using tools, and performing insertions and assembly. End effectors can be mechanical, vacuum or magnetically operated, a snare device, or have some exotic configuration. The design of an end effector is determined by the shapes of objects it has to grasp. Most end effectors are simply some gripping or clamping device.

Control

Control for robots can be a series of stops and limit switches that are tripped by the manipulator durings its motion. They can also be highly complicated computer-controlled devices using machine vision and sonar sensors. With computer-controlled robots, motions of the manipulator and end effector are programmed. In other words, the computer "memorizes" the desired motions. Sensor devices on the manipulator determine the proximity of the end effector to the particular object that is to be manipulated and feed back information to the computer controller for modifications in the trajectory.

FOR HELP

To get you started in building your own robot, the following books, magazines, and suppliers can be used.

Books and Magazines

Several excellent TAB books can be acquired to assist you in developing your own personal robot. They include:

- ☐ *The Complete Handbook of Robotics* by Edward L. Safford, Jr. (No. 1071).
- ☐ *How to Build Your Own Working Robot Pet* by Frank DaCosta (No. 1141).
- ☐ *Build Your Own Working Robot*, by David Heiserman (No. 841).

Other TAB books are also available. Write and ask for a catalog. See also:

- ☐ *Robotica: The Whole Universe Catalog of Robots*, by Texe Marrs and Wanda Marrs

(Stein & Day Publishing, 1895).

A magazine that always provides first-class information and advice to robot hobbyists, including helpful, comprehensive reviews of new robots on the market, is *Robotics Age*. Write to:

Robotics Age
Box 358
Peterborough, NH 03458

Suppliers

Ackerman Digital Systems. Offers the Synthetalker, a speech synthesis board, which uses the Volrax 64 phenome system. Ackerman also sells robotics software.
Ackerman Digital Systems
216 West Stone Court
Villa Park, IL 60181
(312) 530-8992

All Electronics. Offers thousands of surplus electronic parts. Ask for the current catalog.
All Electronics Corporation
905 South Vermont Ave.
P.O. Box 20406
Los Angeles, CA 90006

Altex Electronics. Offers assorted electronics and computer supplies and parts: wire wraps, centronics, gender changers, DIP cables, D.C. mounts, contacts, sockets, and RS-232C converters.
Altex Electronics
618 West Sunset
San Antonio, TX 78216
Toll-free (800) 531-5369
        (512) 828-0503

American Surplus Trading. This company calls itself "the source of electromechanical components for the hobbyist." They have 60,000 items in their inventory, including reconditioned Timex Sinclair 1000 computers, cable assemblies, computer keyboards, switching relay systems, connectors, plugs, and others.

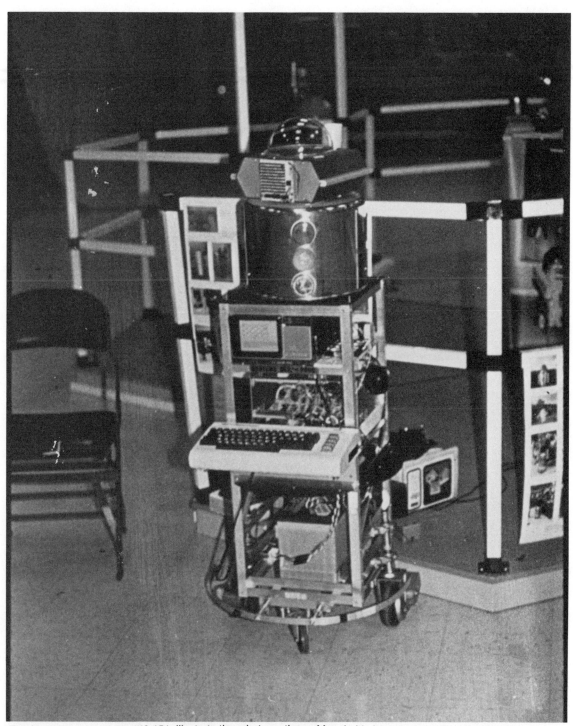

The four photos on pages 149-151, illustrate the robot creations of four hobbyists who presented their home-built robots at a recent robotics show in Albuquerque, New Mexico (courtesy Marilyn Chartrand).

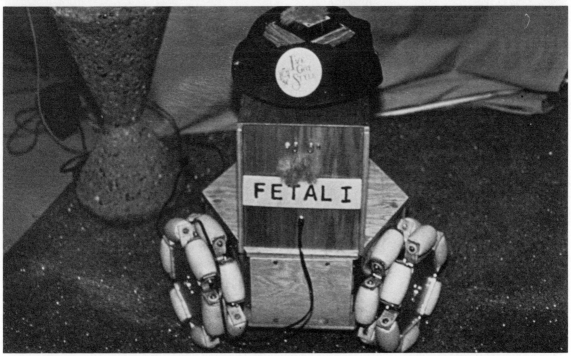

AMERICAN SURPLUS TRADING
62 Joseph St.
Moonachie, NJ 07074
Toll-free   (800) 228-2028
(In Nebraska) (800) 642-8300)

AMSI Corporation. Offers control system parts, including actuators and stepper motors.
AMSI Corporation
Box 651, Smithtown
Long Island, NY 11787
(516) 361-9499

Analog Micro Systems. Offers a wide range of interface software and supplies for educational use; also sells a robotic vision system.
Analog Micro Systems
5660 Valmout Rd.
Boulder, CO 80301

Arctec Systems. Arctec sells a number of outstanding robot products, particularly for the HERO I. These include a HERO Memcom Board that expands the robot's memory to 30K of RAM and an Apple HERO Communicator Board that implements two-way high-speed communication between an Apple computer and a HERO I robot. Also offered: Voice Command System and software packages ("Storyteller," and "Poet"). Storyteller turns HERO into a teller of exciting galactic tales of adventure. For less than $25, the program is highly recommended.
Arctec Systems
9104 Red Branch Rd.
Columbia, MD 21045
(301) 730-1237

Astrotronics Microsystems. Offers the "Big Mouth," a speech synthesizer for robots and computers. The synthesizer has 64 phonemes and inflection and amplitude. It can be programmed to make sound effects and to sing.
Astrotronics Microsystems
1137 Topaz St.
Corona, CA 91720
(714) 734-6006

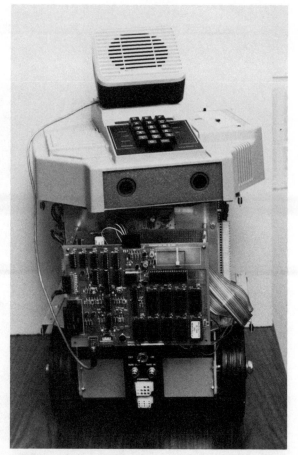

A HERO I robot outfitted with options from Arctec Systems (courtesy Arctec Systems).

Audio Vision. This California company offers a hydraulic robot (about $2,500) which features six controllable movements, positional sensing, and RS-232C computer interface. Also of interest are speech recognition and speech synthesis kits (priced from $59 to $99) for the Apple, TRS 80, ZX 81, VIC 20, Commodore 64, and Timex/Sinclair 1000 computers.
Audio Vision
1279 North Normandie
Los Angeles, CA 90027
(213) 660-5217

B & B Motor and Control. Publishes the 450-page *Motion Control Handbook*, which includes thousands of products, such as motors and controls, electronic and mechanical counters, photoelectric controls, programmable controllers, and stepper motors.

B & B Motor and Control Corporation
Route 4 and Covey Rd.
Burlington, CT 06013
(203) 673-7151

Besearch Information Services. This new company has developed ROBI, an interface for robotics experimenters and hobbyists. The product has four programmable, 8-bit ports, and user-friendly software that transfer files between the computer and the robot.

Besearch Information Services
26160 Edelweiss Circle
Evergreen, CO 80439
(303) 674-0796

John Bell Engineering. Offers six single-board computers and 20 control products used for heat and light control, automatic press control, and robot control.

John Bell Engineering, Inc.
1014 Center St.
San Carlos, CA 94070

Winfred M. Berg. Offers mechanical parts, including gears, shafts, bearings, drive belts, and pulleys. Over 48,500 precision mechanical components in stock. Ask for free catalog.

Winfred M. Berg
499 Ocean Ave.
East Rockaway
Long Island, NY 11518

B. G. Micro. Offers a variety of electronics products for electronic, computer, and robot hobbyists, including controllers, EPROMs, and Z-80 single-board computers.

B. G. Micro
P. O. Box 280298
Dallas, TX 75228
(214) 271-5546

BUYUS. Seven different robot kits are offered, with prices from $25 to $75.

BUYUS
Department RA 185
10 Whitebirch
Ossining, NY 10562

CAL Robot. Markets Heath's Hero-I and RB5X robots and accessories, robot kits, and books.

CAL Robot
P. O. Box 5973
Sherman Oaks, CA 91413
(213) 905-0721

Centroid. Products include stepping motor controllers.

Centroid
Box 739
State College, PA 16804
(814) 237-4535

Computer Parts Galore. This supply house has computer supplies and parts, including ASCII keyboards, 6502 peripheral cards, and parts such as transistors, pin sets, speakers, sockets, connectors, diodes, and inline power jacks.

Computer Parts Galore
56 Harvester Ave.
Batavia, NY 14020
(416) 925-8291

Control Ware. Offers digit I/O board to run motors, lights, and alarms and other products.

Control Ware
Dept. R-3
P.O. Box 1467
Oak Brook, IL 60521
(312) 530-4110

Crabapple Systems. Crabapple Systems has developed an interface card, the KELP, that is designed to interface with Milton Bradley's Robotix 2000 robot construction set. It is for Apple computers and comes with software and documentation. You can also order the Robotix kit from Crabapple. The price of the card is $89.95, while the card plus the kit go for $129.95. For the kit, this is quite a savings over the department store suggested retail price. Shipping and handling charges are $5.00, and Maine residents should add 5 percent sales tax.

Crabapple also offers other computer enhancement products.

Crabapple Systems
118 Commercial St.
Portland, ME 04101
(207) 772-8610

Cyberpak. Stepper motor controller/drivers are what this company offers. Also for sale is a BASIC compatible single board computer.

Cyberpak
P. O. Box 38
Brookfield, IL 60513
(312) 387-0802

Cybot. Cybot has introduced a gripper that is dc-motor driven. At $250, it is ideal for education, research, and hobbyist functions. Mounting is simplified.

Cybot, Inc.
733 Seventh Ave.
Kirkland, WA 98033

Dynetics. Offers robot language translators, communications programs, and computer-robot interface boards. The Colt-I Robot Language Translator allows the user to write easy-to-understand natural language programs on a personal computer and have them translated into HERO robot machine code.

Dynetics, Inc.
306 Wynn Dr.
Huntsville, AL 35805
(205) 837-9234

Edmund Scientific. Offers a variety of electronic and scientific supplies, much of which can be used for robot projects. Items include sensors, parabolic reflectors, gears, springs, and optical components.

Edmund Scientific
101 East Gloucester Pike
Barrington, NJ 08007
(609) 547-8900

Excalibur Technologies. Offers the *Savvy* robotic language for the Apple II computer.

Excalibur Technologies
800 Rio Grande Blvd. NW
21 Mercado
Albuquerque, NM 87105
Toll-free (800) 551-5199
(505) 242-3333

Gloucester Computers. Offers systems for robotic and computer industrial control. The Gloucester systems can even be used to control robot toys. Included is a cartridge that converts the Commodore 64 and VIC 20 into powerful microprocessor development systems.

Gloucester Computer
One Blackburn Center
Gloucester, MA 01930
(617) 283-7719

Herbach & Rademan. Publishes a 32-page catalog of inexpensive parts for robot hobbyists. Innovative parts and supplies include a motorized wheel, sensors, switches, relays, bells and whistles, actuators, and all kinds of motors. The company also offers a Purple Heart box you can award your robot if he or she is wounded in the line of duty—whatever that is.

Herbach & Rademan, Inc.
401 East Erie Ave.
Philadelphia, PA 19134

Frank Hogg Laboratory. In the personal robots section, I reviewed Frank Hogg Laboratory's energetic little NOMAD personal robot. This company also offers a wide variety of exceptional quality parts and supplies. Ask for their current Computer Products Catalog.

The Promqueen/64™ microprocessor development cartridge (top) inserts in the expansion slot of Commodore 64's microcomputer to provide a powerful, low-cost microprocessor development system (courtesy Gloucester Computer, Inc.).

Frank Hogg Laboratory
The Regency Tower, Suite 215
770 James St.
Syracuse, NY 13203
(315) 474-7856

Interface Technology. Offers robot supplies and parts, software, speech recognition devices, and robotics training and education material.
Interface Technology
Box 745
College Park, MD 20740
(301) 490-3608

Jensen Tools. Jensen provides a wide selection of tools, service kits, test equipment, circuit board equipment, computer accessories, and other products suitable for the electronics experimenter or robot hobbyist. The company publishes a free, 160-page catalog describing its products, so ask for it.
Jensen Tools, Inc.
7815 South 46th St.
Phoenix, AZ 85040
(602) 968-6231

Laboratory Microsystems. Robotics control programs in Forth are this company's specialty. Offerings include Forth package for the Z-80, 8086, 68000, and IBM PC systems.
Laboratory Microsystems
P. O. Box 10430
Marina del Rey, CA 90295
(213) 306-7412

The Learning Company. This innovative company's chief robotics product is an educational software package that is a robot construction kit. Designed as an adventure game, *Robot Odyssey I* can help you learn how to design integrated circuits, burn chips, build robots, and improve your logic and thinking skills. Players start by falling into Robotropolis, an underground city of the future inhabited by robots. They can escape only by constructing a robot, which they do in a lab equipped with a tool kit and all necessary instructions. The *Robot Odyssey* package is available for the Apple II family of personal computers. Price is $49.95.
The Learning Company
Dept. EE
545 Middlefield Rd., Suite 170
Menlo Park, CA 94025
(415) 328-5410

Micro Management Systems. Micro Management Systems is a mail-order house that markets discount computers, peripheral computer software, and computer accessories. The company also offers the leading brands of personal robots (RB5X, HERO, etc.), along with software.
Micro Management
2803 Thomasville Rd. East
Cairo, GA 31728
Toll-Free (800) 841-0860

Micromotion, Inc. Offers Master Forth, an assembler language system. A 200-page tutorial and reference manual is included.
Micromotion, Inc.
12077 Wilshire Blvd., Suite 506
Los Angeles, CA 90025
(213) 821-4340

Micron Technology. Offers imaging systems to permit your robot to see.
Micron Technology
1475 Tyrell Lane
Boise, ID 83706
(208) 386-3800

— New Tech Promotions. Distributor to consumers of a number of assembled robots and kits.
New Tech Promotions
2265 Westwood Blvd., Suite 248
Los Angeles, CA 90024

Northwest Computer Algorithms. Sells LISP language systems for robotics artificial intelligence applications. Ask for free catalog.
Northwest Computer Algorithms
P. O. Box 90995
Long Beach, CA 90809
(213) 426-1893

Omega Engineering. Offers thousands of measurement and control sensors and products for computers and robots.
Omega Engineering, Inc.
One Omega Dr., Box 4047
Stamford, CT 06907
(203) 359-RUSH

Perbotics. Offers a 44K programmable memory expansion board for the HERO-I robot, as well as other computer boards.
Perbotics
17072 Emerald Lane #C
Huntington Beach, CA 92647
(714) 847-7846

Polaroid. Polaroid, the same company that brings us cameras, offers a sensitive ultrasonic ranging system that can be used to detect tiny, distant objects and regulate heating and cooling systems and lights. Price is $150.00. Also offered are robot battery packs.
Polaroid Corporation, Battery Division
Mail Stop 4P
Memorial Dr.
Cambridge, MA 02139

Prism Robotics. Offers products for robot hobbyists with the IBM PC personal computer. These products include a Sonar I/O board to facilitate ranging up to 40 feet in distance, a digital interface board, and stepper motor controller/driver boards.
Prism Robotics, Inc.
P. O. Box 9474
Knoxville, TN 37920
(615) 573-4944

Remote Measurement Systems. Distributes remote sensors such as those that measure wind speed, humidity, and light levels.
Remote Measurement Systems
P. O. Box 15544
Seattle, WA 98115
(206) 525-3369

Rio Grande Robotics. According to this company, it is the first to offer a full range of robots, kits, software, and languages. Rio Grande is your one-stop store for robotics. Here you can buy MARVIN, RB5X, turtles, robot shop kits, Rhino Robots products, and much more, but no parts.
Rio Grande Robotics
1595 West Picacho #28
Las Cruces, NM 88005
(505) 524-9480

Robitech. Offers affordable modular components and controls for robots and robotic systems.
ROBITECH
10 Upton Dr.
Wilmington, MA 01887
(617) 657-6143

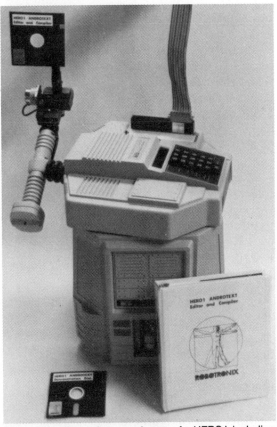

Robotronix offers software and games for HERO I, including ANDROTEXT, a high-level robot language (courtesy Robotronix).

Robohand. Give your robot a "smart hand," a tactile (touch) sensing device. This company has a whole line of robotic grippers.
Robohand, Inc.
P. O. Box 438
Easton, CT 06612
(203) 374-6063

Robotronix. An excellent source of software for the HERO robots. Among the packages offered: HERO Butler ($25), HERO Patrol ($25), HERO Pet ($25), and several games (voice control), such as Hop-to-it, (Math Quiz), which go for only $15 each. Add $3 for shipping and handling. Robotronix also has a high-level language system for robots called Androtext™.
Robotronix, Inc.
Box 1125
Los Alamos, NM 87544

Schulz Enterprises. Offers single board computers, various control and interface boards, and voice syntheses devices.
Schulz Enterprises, Inc.
1285 Las Tunas Dr.
San Gabriel, CA 91776
(818) 287-5067

Small Parts. Offers mechanical parts and components for robot hobbyists, including gears, bearings, pulleys, and tools.
Small Parts, Inc.
6901 North East Third Ave.
P. O. Box 381736 R
Miami, FL 33138
(305) 751-0856

Software Science. Offers sensors, power supplies, and accessories.
Software Science
P. O. Box 44232
Cincinnati, OH 45244
(513) 561-2060

Spectron Instrument. Offers mobile robot kits, software, courseware, vision systems, and some parts.

Spectron Instrument
1342 West Cedar Ave.
Denver, CO 80223
(303) 744-7088

Stock Drive Products. This company offers parts, parts, and more parts—on-the-shelf and ready to use. Also, the company markets robot kits (see Part II, Stock Model Products). Stock Drive has two excellent (and comprehensive!) catalog/manuals that are highly recommended: *Design and Application of Small Standardized Components*, Data Book #757 (includes a 51-page robotics design section) and *Handbook of Small Standardized Components*, Master Catalog #757.
Stock Drive Products
55 South Denton Ave.
New Hyde Park, NY 11040
(516) 328-0200

Street Electronics. The Echo Speech Board produced by this firm is a text-to-speech system that creates a pleasant and natural-sounding female voice. It is compatible with any digital system with an RS-232C interface. Users send standard ASCII characters to generate unlimited speech. The system has 63 pitch levels and extremely accurate pronunciation.
Street Electronics Corporation
1140 Mark Ave.
Carpinteria, CA 93013
(805) 684-4593

Dr. Clyde T. Suttle (Meccano Systems). Offers a large selection of ready parts for robot builders.
Dr. Clyde T. Suttle
6062 Cerulean Ave.
Garden Grove, CA 92645

Tech-O. Logic stepping motor controllers and transistors.
Tech-O
P. O. Box 176
Los Alamitos, CA 90720
(213) 596-3677

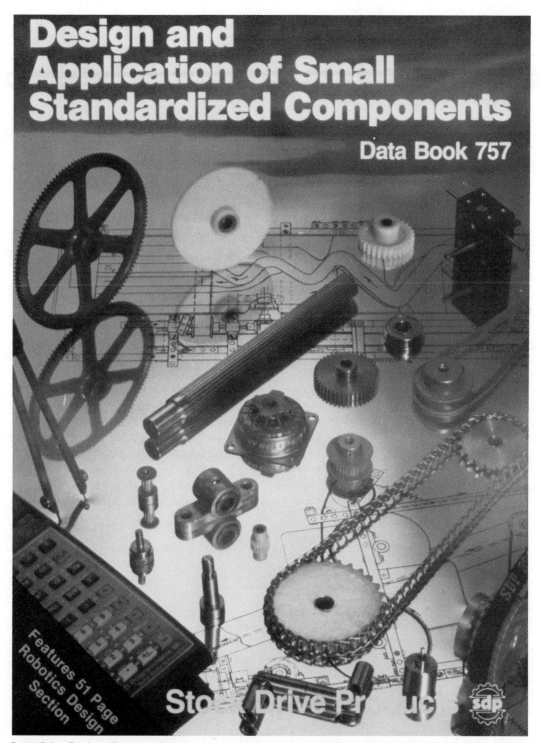

Design and Application of Small Standardized Components

Data Book 757

Features 51 Page Robotics Design Section

Stock Drive Products

Stock Drive Products has an exhaustive inventory of parts for the robot builder (courtesy Stock Drive Products).

Thrifty Bits. Offers RB and Androbot robots at a discount, plus electronic testing equipment.
Thrifty Bits
2722 Windy Bush Rd.
Newtown, PA 18940
(215) 598-8119

TransTech Systems. Offers unique tabletop kits which model robotics and manufacturing technology systems. These are excellent creative learning tools.
TransTech Systems Division
Dept. RA 11-84
9889 Hibert St., Suite E
San Diego, CA 92131
(619) 566-2882

Tri-Tronics. Offers fiber optic sensors for robot vision.
Tri-Tronics, Inc.
Tampa, FL
Toll-free (800) 237-0946

2500 A.D. Software. Assemblers CZ-80, 8086/88, and Unix (plus other systems) and cross-assemblers.
2500 A.D. Software
P. O. Box 4957
Englewood, CO 80155
(303) 790-2588

Vantec. Offers computer and robot controller to command speed and direction of two independent motors from a single parallel output port.
Vantec
15445 Ventura Blvd., Suite 10-281
Sherman Oaks, CA 91413
(818) 993-1073

Vernitech. This New York firm specializes in laser scanning motors, stepper motors, encoders, clutches, and brakes.
Vernitech Corporation
300 Marcus Blvd.
Deer Park, NY 11729

Virtual Devices. Offers remote control and communication systems for operation of the HERO I, letting you control the robot from your personal computer.
Virtual Devices
Box 30440
Bethesda, MD 20814
(800) 762-7626

Voice Machine Communications. Offers voice recognition equipment and voice-controlled word processing software that allows handicapped persons such as quadriplegics to compose and print written communication.
Voice Machine Communications
Division of Kinetics Technology International
1000 South Grand Ave.
Santa Ana, CA 92705
(714) 541-0454

Appendices

Personal Robot Hobbyist Organizations

Several organizations have sprung up to support the growing interest in personal robotics. A concise description of the major groups follows. I apologize to any group inadvertently left out. If your group is not mentioned, please send information about it to me, c/o TAB Books, Inc., Blue Ridge Summit, PA, and it will be included in the next edition of the book.

ROBIG: ROBOTICS INTEREST GROUP

ROBIG is a Virginia group. Its members include persons building their own robots for science fairs, robotics educators, and others. Dues are $15. Write to:
Robotics Interest Group (ROBIG)
3205 Sydenham St.
Fairfax, VA 22021

RSA: ROBOTICS SOCIETY OF AMERICA

The RSA is a national group with chapters in many states. The organization is active on a number of fronts. Regular chapter meetings are conducted where guest speakers present robotic models or discuss important issues about the field. A periodic newsletter keeps members advised of goings-on around the world in the realm of personal and industrial robots. The newsletter also contains an informative calendar of events which lists important robot conventions, exhibits, trade shows, and other activities around the nation.

The purpose of RSA is three-fold:

- ☐ To provide support, both moral and technical, to all people who are engaged in or are ready to begin developing their own thinking machines;
- ☐ To help promote a beneficial relationship between human and constructed intelligence;
- ☐ To encourage natural curiosity and spread knowledge about robotics among young people and the general public.

Annual dues for RSA membership are $15 for students, $25 for personal, $35 for family, and $100 corporate. This group is an excellent one, and if you don't have a local chapter, you may even consider establishing one. Write to RSA for a list of local chapters and for additional information.
Dr. Walter Tunick, Executive Director
Robotics Society of America
200 California Ave., Suite 25
Palo Alto, CA 94306

HOMEBREW ROBOTICS CLUB

The Homebrew Robotics Club has several active chapters. Members meet monthly to discuss new products and advances in robotics and review ongoing robotics projects. Technical subjects are often addressed in informative presentations—subjects like voice synthesis, grippers, increasing robot intelligence, and enhancing mobility. Many of the greatest achievements in personal computer history emanated from a homebrew computers group. The same may well prove to be true in robotics. For more information, contact:
Richard Prather
91 Roosevelt Circle
Palo Alto, CA

REAL: ROBOTICS EXPERIMENTERS AMATEUR LEAGUE

REAL members are dedicated hobbyists and experimenters who believe in the future of robotics and cross-feed useful information about robotics research and development. Many members have built their own robots or are working on such projects. This is a nationwide organization with groups in California, Georgia, and Massachusetts, and others forming elsewhere. For more information, contact:
Robotics Experimenters Amateur League
P.O. Box 3227
Seal Beach, CA 90740

UNITED STATES ROBOTICS SOCIETY

This group is for the "garage experimenter." Members receive a monthly newsletter, bulletins, and discounts on books and robot supplies.
United States Robotics Society
616 University Ave.
Palo Alto, CA 94301

OTHER ROBOTICS ORGANIZATIONS

Although they are not specifically for personal robot hobbyists, two other groups deserve mention. One is a society for professionals working in the robotics industry; the other is a trade group.

Robotics International/Society of Manufacturing Engineers

The Society of Manufacturing Engineers (SME) is an international organization dedicated to the continuing education of engineers, technologists, and others. SME's Robotics International division is extremely effective, sponsoring many first-rate conferences, symposia, and clinics. This group publishes the outstanding periodic publication, *Robotics Today* (6 times per year, subscription rate $7.50 for members; $36 for nonmembers). For a subscription or for additional information, write or phone:
Robotics International/SME
One SME Dr.
P.O. Box 930
Dearborn, MI 48128
(313) 271-1500

Robotics Industries Association

The Robotics Industries Association (RIA) is a trade association serving the robotics field. Members include all types of companies: those making industrial robots, producers of personal robots, robotics consultants, distributors, users, researchers, and others. RIA provides a newsletter to its members and sponsors trade shows and expositions, market studies, workshops, and many other activities. I have had the opportunity of meeting the personnel who administer this organization and was impressed with their energy, dedication, and excep-

tional work in the field of robotics. For more information, write or phone:
Robotics Institute Association
One SME Dr.
P.O. Box 930
Dearborn, MI 48128
(313) 271-0778

Appendix B

Manufactures

Following is a list of manufacturers, their addresses, phone numbers (where given), and some of their products.

Acorn Computers Corp.
400 Unicorn Park Dr.
Woburn, MA 01801
617-935-1190
Cyber 310 arm robot

Amarobot
1780 Shattuck Ave.
Berkeley, CA 94709
415-548-3554
MARTHA, GEORGE, TURTLE, ISAAC, HYDROBOT

American Micro Products
705 N. Bowser
Richardson, TX 78081
GR-15 robot plotter

Androbot, Inc.
101 E. Daggett Dr.
San Jose, CA 95134
FRED, BOB, TOPO III

Android Amusement Corporation
1408 East Arrow Hwy.
Irwindale, CA 91706
213-303-2434
ANDREA, ADAM, DC-2

Arctec Systems, Inc.
9104 Red Branch Rd.
Columbia, MD 21045
301-730-1237
Gemini

Axlon
P.O. Box 306
125 Main St.
Half Moon Bay, CA 94109
408-745-1110
ANDY

Chessel Microproducts
Southdownview Rd., Worthing
West Sussex BN 14 8NL
England
GR-15 robot plotter

ComRo Inc.
126 East 64th St.
New York, NY 10021
202-751-7414
ComRo ToT

Cyberotics, Inc.
99 West St.
Medfield, MA 02052
617-359-3344
CYBER I

D & M Computing
Box 2102
Fargo, ND 58107
ARMDROID I, II, ZEAKER robot arms

Eastern Machinery & Manufacturing
Robotics Division
850 South West Temple
Salt Lake City, UT 84101
801-532-5430
Q-bot kit

Economatics, Ltd.
4 Orgreave Crescent
Dore House Industrial Estate
Handsworth, Sheffield
England
BBC Buggy

EZ Mower, Inc.
P.O. Box 77018
Houston, TX 77215
ROBOMOWER

Feedback, Inc.
620 Springfield Ave.
Berkeley Heights, NJ 07922
Toll-Free 800-526-8783
NJ 201-464-5181
ARMOVER, ARMADILLO

Frank Hogg Laboratory
770 James St.
Syracuse, NY 13203
315-474-7856
NOMAD Radio Shack

Genesis Computer Corp.
P.O. Box 152
Hellertown, PA 18055
215-861-0850
NOMAD C-64, VIC 20

Harvard Associates
260 Beacon St.
Somerset, MA 02143
617-492-0660
617-492-2999
Robotnick P/L Robot Arm, Tassman Turtle, Turtle Tot

Heath Co.
Dept. 150-375
Benton Harbor, MI 49022
Toll-Free 800-253-0570
AL, MI, HI 616-982-3411
Catalogs free

Heathkit/Zenith Educational Systems Div.
Veritechnology Electronics Corp.
Dept. 570-445
St. Joseph, MI 49085
616-982-3206
Schools

Hob-Bots of Hudson
P.O. Box 640
Hudson, OH 44236
216-656-3413
Micro Orvoid, Sentroid

Hobby Robot Co.
P.O. Box 887
Hazelhurst, GA 31539
or
Micro Management Systems, Inc.
2803 Thomasville Rd. East
Cairo, GA 31728
912-377-7170
SMART RABBIT

Hubotics
6352-D Corte del Abeto
Carlsbad, CA 92008
619-438-9028
HuBot

International Robotics
611 Broadway, Suite 422-B
New York, NY 10012
212-982-8001
SICO

Iowa Precision Robotics, Inc.
201 Main St.
Melvin, IA 51350
Toll-Free 800-831-7927
712-736-2600
MARVIN MARK I

Lab-Volt System
P.O. Box 686
Farmingdale, NJ 07727
Robot arms

Microbot, Inc.
453-H Ravendale Dr.
Mountain View, CA 94043
703-455-5088
TeachMover robot arm

Personal Robotics Corp.
469 Waskow Dr.
San Jose, CA 95123
408-281-7648
RoPet

Prep, Inc.
1007 Whitehead Rd. Ext.
Trenton, NJ 08683
Toll-Free 800-257-5234
609-882-2668
SCORBOT-ER III robot arm

RB Robot Corp.
18301 W. 10th Ave.
Suite 310
Golden, CO 80401
303-279-5525
RB5X

Remcon Electronics
P.O. Box 148258
Nashville, TN 37214
615-361-3936
Teach Robot robot arm

Rhino Robots, Inc.
P.O. Box 4010
Champaign, IL 61820
217-352-8485
SCORPION

ROBOFORCR Magazine
300 Madison Ave.
New York, NY 10017

Robot Repair
816 1/2 21st St.
Sacramento, CA 95814
916-441-1166
ROBOCYCLE

Robot Shop
P.O. Box 582
El Toro, CA 92630
Robot Shop kit or product

Robotex
111 East Alton
Santa Ana, CA 92707
714-556-8679
RM-501 robot, RTROL language

Robotland
4251 N. Federal Hwy.
Boca Raton, FL 33431
305-368-8118
Elami Robots

Servitron Robots, Inc.
1009 Grant St.
Denver, CO 80203
Toll-Free 800-457-0001
AL, CO, HI 303-831-9370

ShowAmerica, Inc.
841 N. Addison Ave.
Elmhurst, IL 60126
312-834-7500
Quadracon, Peeper

Stock Model Parts
54 S. Denton Ave.
New Hyde Park, NY 11040
516-328-3333
LINE TRACER II, MEMOCON CRAWLER

Technical Micro Systems, Inc.
P.O. Box 7227
Ann Arbor, MI 48107
313-994-0784
ITSABOX

The Robot Factory
P.O. Box 112
Cascade, CO 80809
303-687-6208
COMMANDER ROBOT

Tomy Corp.
901 East 233 St.
Carson, CA 90749
OMNIBOT, VERBOT, DINGBOT

Glossary

Glossary

This glossary will help you to understand many of the terms, acronyms, and abbreviations used in this book that are unfamiliar to you. If you are interested in a more complete robotics glossary, the Robotics Industries Association (RIA) offers its *RIA Robotics Glossary* for $8 postpaid. Write to the RIA at P.O. Box 1366, Dearborn, MI 48121.

actuator—A device (motor or transducer) that converts electrical, pneumatic, or hydraulic energy to place a robot in motion.

algorithm—A process, prescribed set of rules, or mathematical equation to solve a problem in a finite number of steps.

android—A robot closely resembling a human in its physical appearance. Androids are not necessarily equivalent to humans in shape; for example, wheels may suffice for human legs and feet.

anthropomorphic—A robot or robot with human shape and joints whose movements are similar to those of humans.

artificial intelligence (AI)—Man-created intelligence comprised of human intelligence functions such as logical reasoning, induction, deduction, thinking, adaptation, and self-correction. As with human (real) intelligence, there are many degrees, levels, types, and functions of artificial intelligence.

axis—A path in space in which an object travels in any of three dimensional directions: X (parallel to the earth), Y (parallel to the earth and perpendicular to "X"), and Z (vertical and perpendicular to the earth's surface).

bit—Stands for "binary digit." The smallest amount of information a computer can hold.

byte—Eight bits make up one byte. A byte is a unified data storage unit accepted by a computer's memory.

cassette recorder—A computer or robot peripheral that allows the transfer of information between the machine's memory and the magnetic tape run by the cassette recorder. The magnetic tape program directs the actions of the computer or robot.

central processing unit (CPU)—The control, or hardware, part of a computer (command center) in which microprocessor(s) direct the computer's sequence of operations and executions. Also called the operating system.

controller—A device for communication that a person uses to send commands to the control system of a robot. A controller may be on-board or remote from the robot.

CYBORG—Stands for CYBernetic ORGanism. A cyborg, often called a bionic man or woman, is usually a human modified by the addition of artificial organs and limbs. A *cyberman* also has a human brain. A *cybot* is a mechanical man with a humanlike brain. A cyborg, then, is part biological, part electromechanical.

debug—Find and correct (troubleshoot) mistakes or errors (bugs) in a computer or robot program.

documentation—A reference manual, or guide, to the operation and functioning of a software program, computer, robot, or peripheral device.

downloading—The sending, or transmission, of a computer program by electronic means.

end-effector—The hand or claw of a robot arm. Also called an actuator, gripper, or manipulator because it grasps and acts upon objects.

hardware—The physical apparatus of a robot or computer, consisting of the internal and external electrical, pneumatic, and hydraulic parts of the robot and its peripheral devices.

hydraulic motor—A system (actuator) of values and pistons or vanes which convert high-pressure fluid into mechanical rotation.

I/O—Abbreviation for input/output.

kilobyte (K)—1,024 bytes of memory.

language—A set of symbols and rules for communicating between people, computers, and robots. Popular and functional robot languages include Androtext, Forth, Fortran, and BASIC.

LED display—Light-emitting diode characters comprising an illuminated visual readout.

microprocessor—A section of integrated circuit chips which direct computer operations.

non-volatile memory—Memory a computer retains when its power supply is cut off.

pick-and-place robot—This type of robot is limited. Its simple operation consists of picking up and transporting items, such as parts, from point-to-point.

programmable—A computer, including that of a robot, which can be instructed, or commanded, to operate in a specific, programmed manner or perform certain, directed tasks.

random access memory (RAM)—The memory used to store data, or a program, in a computer or robot.

read only memory (ROM)—The permanently stored memory base of the computer or robot.

RS-232C—A connecting device or medium (called a "port") which permits the interfacing of robots and computers.

sensor—An input/output feedback device that senses (recognizes or interprets) activity. Sensors permit the robot to "read" and react to its environment. *Contact* sensors sense mechanical contact; *optical* sensors convert light into recognizable electrical signals; *sonic* sensors convert sound into recognizable electrical signals. Intelligent robots take actions based on their sensory perceptions.

servo-motors—Motors that stop and move ("drive") a robot's physical features.

software—The program of prescribed sequences that controls the operation of a computer or robot.

teach pendant—A handheld remote control unit

with which a robot's actions can be commanded or programmed.

transducer—A sensory device that converts physical attributes or parameters (water or air pressure, temperature, weight, etc.) into electrical signals.

TTY—Abbreviation for teletype.

uploading—The receiving of a program.

utility—A program designed to help and assist a computer programmer.

volatile memory—Computer memory that is lost when power is cut off.

Index

Index

OTHER POPULAR TAB BOOKS OF INTEREST

TAB | TAB BOOKS Inc.

Blue Ridge Summit. Pa. 17214

Send for FREE TAB Catalog describing over 750 current titles in print.